DATE DUE

DEMCO 38-296

A SHEARWATER BOOK

RE**Q**UIEM
FOR
NA**T**URE

JOHN TERBORGH

RE QUIEM
FOR
NATURE

ISLAND PRESS / Shearwater Books

Washington • Covelo • London

A Shearwater Book
published by Island Press

Copyright © 1999 John Terborgh
Preface to the paperback edition copyright © 2004 John Terborgh

First Island Press cloth edition, April 1999
First Island Press paperback edition, July 2004

Library of Congress Cataloging-in-Publication Data
Terborgh, John, 1936–
    Requiem for nature / John Terborgh.
        p.   cm.
    Includes bibliographical references.
    ISBN 1–55963–587–8 (cloth: alk. paper) ISBN 1–55963–588–6 (pbk: alk. paper)
        1. Nature conservation—Tropics.   2. Protected areas—Tropics.
    I. Title.
    QH77.T78T47   1999   99–21385
    333.7'2'0913—dc21   CIP

Printed on recycled, acid-free paper

Manufactured in the United States of America

10   9   8   7   6   5   4   3   2   1

CONTENTS

Preface to the
Paperback Edition

ALMOST SIX YEARS have elapsed since the
hardcover edition of *Requiem for Nature* went to press. Much has happened during that interval as the pace of global change has continued
to accelerate. The occasion of this paperback edition provides an opportunity to take stock: Have the predictions I made in the late 1990s
been supported by subsequent developments? The answer is yes and
no, and the news is both good and bad.

One prediction was of continued deterioration of the conservation
situation in West Africa, a prediction that has proven distressingly accurate. Côte d'Ivoire (Ivory Coast), Sierra Leone, and Liberia have
been convulsed by civil wars, and the political situation in all three
countries remains precarious. Illegal logging and mining have financed
the various warring factions, to the detriment of wildlife and parks.

A study conducted in Ghana, one of the few peaceful countries in
the West African forest region, provides an unsettling perspective on
the future of wildlife in that part of the world. Shortly after attaining
independence in 1960, Ghana established six national parks to represent the country's varied ecosystems. At the time they were established, the six parks harbored forty-one large mammal species, including lion, elephant, buffalo, leopard, hippo, antelope—in short, the

whole panoply of charismatic African animals. But by 1998, eleven species had disappeared from all six parks (and presumably from the entire country), and another eleven hung on in only one park. Only the baboon, an animal that adapts well to human presence, remains in all six Ghanaian parks. The problem: poaching. The parks all have guards, but they are too few and too poorly equipped to oppose legions of furtive hunters that sneak across long, uncontrollable borders. These findings should send a strong message to conservation organizations: if poaching cannot be controlled in West African parks, the distinctive wildlife of that region—pigmy hippo, Diana monkey, Miss Waldron's red colobus monkey, and others—will not long remain in the wild.

In an unanticipated conservation disaster, the lowland forests of Indonesia have been ravaged in a rapacious wave of logging in the few years since the sudden collapse of the Suharto regime in 1998. Indonesia is a megadiversity country harboring vast numbers of endemic plants and animals. Sudden relaxation of the ironfisted authority maintained by Suharto's central government unleashed a free-for-all on Indonesia's remaining primary forests, including those in national parks, as social order disintegrated in many parts of the geographically scattered and ethnically diverse nation. A succession of weak governments has proved unable to reassert central authority and has yielded to pressure to transfer administrative control over provincial lands to corrupt local officials, ensuring further degradation of remaining forests.

Now for some good news. When *Requiem for Nature* was written in the late 1990s, the global trend in creation of new protected areas was strongly downward, making it appear as if the era of park creation was nearing its end. In the face of growing human populations and increasing pressures on natural resources, it appeared that many governments viewed the opportunity cost of protecting more habitat as prohibitive. But fortunately that judgment proved premature. The nations of the earth seem to have acquired a second wind for park creation that has yet to run its course.

Thus, the 3,000 delegates to the Fifth World Parks Congress held in Durban, South Africa, cheered at the announcement that as of 2003, 11 percent of the earth's terrestrial realm had been incorporated into formal protected areas. Even more impressive is the trend it signifies: the 2003 total represents a doubling since the previous congress in 1993 and a tripling since the 1982 congress, affirming that nature conservation still enjoys a high place on the world's agenda.

The best news comes from Latin America and Africa where some countries are making truly impressive commitments to conserving nature. Venezuela, the best case example, has protected 46 percent of its national territory, 17 percent of it in national parks; Panama has protected 33 percent and the Dominican Republic 32 percent; a number of other countries in the region are not far behind. On a subnational level, the Brazilian state of Amapá may lead the world. In 2001, Brazil inaugurated the 5-million-hectare Tumucumaque National Park in Amapá, and in 2003, the state government announced the creation of six additional state-administered reserves, bringing the total to 72 percent of the state's territory. Such thrilling demonstrations of leadership and political will set a high bar for other nations.

There has been dramatic progress in Africa as well. Giant multinational transboundary megaparks have been declared in both central and southern Africa. And in Madagascar, President Marc Ravalomanana has committed his government to increasing the area under protection from 3 percent to 10 percent of the national territory, an amount sufficient to include all natural vegetation remaining on the island.

Countries all over the world are taking seriously their moral responsibility to conserve nature, as attested by their continuing willingness to expand protected areas. The 11 percent global total is still not enough, but attitudes nearly everywhere are changing in a positive direction and the political will to create additional nature parks has not yet been exhausted. If the momentum on display at the Durban World Parks Congress can be sustained for a few more decades, a brighter picture will emerge.

Meanwhile, new challenges have come into focus since the publication of the hardcover edition of *Requiem for Nature*. The most acute are the bush meat trade and illegal logging. Explosion of the bush meat trade in central Africa is decimating gorillas and chimps and dozens of other species, and threatens to empty African forests of their animal inhabitants. A similar story can be told of freshwater turtles in Southeast Asia. Controlling the commercial trade in wild animals is fundamentally a matter of political will. Kenya, for example, successfully prohibited the commercialization of wild game decades ago. Other African and Asian countries should be pressured to adopt similar policies.

Illegal logging has become one of the gravest threats to the survival of forests and their wildlife in nearly every corner of the tropical world. The pernicious wild meat trade is intimately linked to logging because loggers construct the roads that provide access to hunters. Since developed countries provide much of the market for illegally harvested timber, all nations bear a collective responsibility for curtailing illegal trade.

International pressure can work. A decade ago elephants, rhinos, tigers, and bears were in dramatic decline when their body parts commanded high prices in global markets. Illegal traffic in animal parts persists, but on a greatly reduced scale. Meanwhile, populations of at least some persecuted species—African elephants, black rhinos, and bears (in the United States)—are recovering. Success in reducing the trade in animal parts is attributable to the effectiveness of high-profile publicity campaigns orchestrated by the World Wildlife Fund and other organizations. Similar campaigns now need to publicize the devastating impacts of the bush meat trade and illegal logging.

The above news items are but a sampling of recent developments on the conservation front. There is good news and bad, progress and regression. From the perspective of the moment, it is difficult to extract the signal from the noise because the turbulence of change is marked by contradictions, as competing interests and philosophies push and shove in the quest for primacy. Most contemporary events have their antecedents and their logic, and that is true here. *Requiem for Nature* offers a window of understanding into the pressures and trends that underlie today's news. The book explores the tensions between exploitation and sustainable development, between the aspirations of the rich and those of the poor, and between traditional lifestyles and globalization. How these tensions are resolved will determine whether nature survives.

Humanity is slowly reaching a consensus that the human species occupies a commons called planet Earth for which all of us bear collective responsibility. Neither the rich nor the poor can be exempted. The haves need to provide assistance, both technical and financial, to the have-nots. All humanity must join in a mutual effort to ensure the survival of a healthy earth. I sense that good will is there at the grassroots level. What is currently most lacking is committed leadership at the political level. A little leadership at the highest level could go a long way toward forestalling the day when the bell tolls requiem for nature.

John Terborgh
March 2004

Preface

FROM MY EARLIEST days as a preschooler, my abiding obsession has been discovering and appreciating nature. Among the fondest memories I retain from childhood are of summer days roaming the woods and streams around my family's northern Virginia home. My goal was to discover all the snakes, turtles, salamanders, birds, mammals, and what-have-you that inhabited these surrounds. The thrill of each new find contributed to a budding knowledge of natural history. Further quests were inspired by the many field guides I memorized, for the diversity of nature was an endless source of fascination and of challenges yet unmet.

Those were idyllic times, brought to a harsh and abrupt end after the close of World War II, when the bucolic tranquility of our neighborhood was shattered by an onslaught of bulldozers, trucks, and construction workers. Seemingly overnight, the forests vanished. Broad red scars in the earth, the future streets and lanes of middle-class suburbs, bespoke the agony of the land. For me, the experience was shattering. The forest that had been the source of my greatest satisfactions was destroyed in front of my eyes, and I was powerless to do anything about it. A sense of personal violation produced a deep outrage that lingers today. But now I am not quite so powerless as I was then. I can write, which at least allows me to make my anguish known.

Nearly fifty years have passed since bulldozers despoiled my cherished childhood haunts. Wherever I have gone during those years, I have seen the same process repeated. Around the world, the beauty of nature is being replaced by the banal handiwork of humans. But to what end? Does the next sprawling development being built just beyond where the last ends add measurably to anyone's quality of life? No, to the contrary, the quality of life ratchets downward each time a bit of countryside is transformed into yet another row of garish fast-food drive-ins, motels, and automobile dealerships. A once relaxing and eye-pleasing parcel of land might then be better described as jarring, stress inducing, and unsightly. And yet the march of development goes on and on, with no end in sight. When I let my mind wander and try to imagine the next fifty years as an extension of the past half century, I wonder whether the mental picture foretells a better world for anyone. Certainly there will be a few more rich developers and car dealers than today, but for the vast majority, the benefits of endless demographic and economic expansion are not apparent. For most of us, the added congestion, pollution, and isolation from nature represent decrements, not increments, in the enjoyment of our lives.

As a scientist concerned for the welfare of all life, I view these trends as highly threatening to the future of both humankind and nature. No country on earth is entirely sheltered from the relentless press of forces that promise to roll over what is left of the natural world. Some would argue otherwise, pointing to institutions that have been created by humans to preserve nature for posterity. We call them parks. Nature is safe in parks, or so it is presumed. But that is a blithe presumption, made without reference to nature's requirements and with scant knowledge of the forces that have perpetuated the earth's vast biological wealth over past eons. As scientists begin to understand how nature works and to identify the forces that ensure nature's continuity, it is becoming painfully apparent that many parks are not up to the job assigned to them. That is one part of the story I tell. The other has to do with the limitations of parks as human institutions and what might be done to overcome these limitations.

In the United States, parks are robust institutions, protected from degradation and backed by strong, if not always solvent, government agencies. If parks were equally robust in the Tropics, where most of the earth's biological wealth resides, I might have cause for complacency. But what I have seen in a lifetime of roaming the Tropics in quest of unspoiled nature has dispelled any such complacency and replaced it with panic. What I see are inadequate parks, unstable societies, and faltering institutions. Nature cannot long survive any of these debilities; all three in combination add up to a recipe for extinction.

Nature can be saved, but only through a thoughtful combining of good science and strong institutions. Right now, much of the world benefits from neither. We have a long way to go before anyone can feel comfortable about the future of nature. And there is no time to lose in getting on with the journey.

Acknowledgments

My thinking about many topics has been sharpened and focused through conversations with friends and colleagues. Among these, I am particularly indebted to Donald Brightsmith, Lisa Davenport, Deborah Lawrence, Carlos Peres, Laura Snook, Thomas Struhsaker, Carel van Schaik, and Douglas Yu. All read and commented on parts of the manuscript. Lisa Davenport and Douglas Yu each read the entire manuscript, parts of it more than once, and offered many constructive suggestions.

I might never have embarked on this undertaking were it not for the wonderful boosts I received as a Pew Conservation Scholar and as a MacArthur Fellow. The support accompanying these two honors carried me through a long funding drought and kept my spirits strong when otherwise I might have lost heart.

Both the first and second drafts of the manuscript were written at my tropical hideaway in Perú's Manu National Park. The Manu is truly my second home, and for the privilege of being able to spend several months each year in the Garden of Eden, I am everlastingly grateful to the park's administration and to the Instituto Nacional de Recursos Naturales (INRENA), Perú's central authority for national parks.

My editor, Laurie Burnham of Shearwater Books, had the insight to see when I was off track and to point me in more productive directions. Such astute editors have been rare in my experience, and I remain both impressed and grateful.

Finally, I warmly thank my long-suffering assistant, Marty Jarrell, who has to put up with the idiosyncrasies of a harried boss who abandons her to solitude for five months a year while he immerses himself in tropical nature.

J. T.
Cocha Cashu Biological Station

REQUIEM
FOR
NATURE

CHAPTER 1

THE MAKING
OF A DISSIDENT

WHEN I FIRST laid eyes on Perú's Apurímac valley, I declared it to be the most beautiful place I had ever seen. The sinuous Apurímac River sparkled in the bright sunshine, radiating an intense blue color as it coursed swiftly over beds of sand and gravel between towering walls of virgin forest. At close range, the passing river emitted an audible hiss, the sound of roiling gravel being swept along the bottom by a powerful current.

Soaring ranges of the Andes frame both sides of the fertile valley through which the Apurímac flows. Tier upon tier of forest-clad ridges mount into the cloud-shrouded distance, creating extravagant vistas of wild inaccessibility that I found irresistibly alluring. For me, a tropical biologist at the beginning of my career, to be in the heart of such a wholly pristine scene was both sublime and exhilarating.

More than thirty years have passed since I began a career of studying tropical forests in the Apurímac valley. The valley then had few inhabitants, but those few had already irrevocably altered the status quo. These self-proclaimed pioneers had claimed the best sites in the valley. The indigenous Campa Indians, once sovereign over the whole

3

region, had been displaced from the fertile valley floor and had re-treated into the foothills.

To the Campas, the pioneers were alien invaders and usurpers. To the pioneers, Indians didn't count. By the criteria of the pioneers, the forest was unclaimed wilderness because it had not been cleared. Ownership could be claimed only of cleared land on which someone had obviously toiled. Everything else was up for grabs.

To be sure, the first pioneers were intrepid souls, for reaching the valley in those days entailed a grueling seven-day trek over the Andes with a train of pack animals. These first settlers to breach the wilder-ness were driven not by a love of solitude and nature but by the lure of fertile land and what it promised for the future. These were ambitious people, determined to create wealth out of virgin nature.

The completion of a road into the valley brought the wilderness idyll to an abrupt and jolting end. Financed by the Alliance for Prog-ress (Alianza para el Progreso), President John F. Kennedy's much ballyhooed international economic development program, the road opened the floodgates to a second wave of invaders. These were not the scions of wealthy landowning families, as the first pioneers had been, but landless and illiterate peasants from the Andean highlands. In the vanguard of a demographic explosion that continues to drive people into rain forests the world over, these *colonos* (settlers) verita-bly poured into the valley. Every arriving truck carried several families perched atop the load. With little more than their clothing, an ax, and a machete, they set off into the forest in search of a dream, land they could call their own. Many of them had never seen a tropical forest before.

With the weekly arrival of scores of families encouraged by a gov-ernment-sponsored land distribution program, the frontier melted away in what was, in retrospect, no more than the blink of an eye. An impromptu shantytown sprang up where the road ended on the riverbank. The once proud but now demoralized Campas retreated into the distance, always one jump ahead of the *colonos*.

By 1972, the year I last saw the Apurímac valley, the population of *colonos* had swollen to more than a hundred thousand and was still growing steadily. By then, hardly a tree remained of the magnificent forest that had so recently filled the valley bottom. Plantations of coffee, cacao, and coca and slash-and-burn patches had replaced the forest and were appearing on the lower slopes of the mountains, a sign that the fertile valley floor had all been claimed.

When I decided to leave the valley, I knew I would not return. I left because the wild nature that had drawn me there had been extinguished in just seven years. What has happened to the Apurímac valley in the years since my last visit concentrates in one small region all the passions and violence that frontier zones inspire. The soils and climate of the valley are ideally suited to the cultivation of coca. Coca was widely grown there during my time, but it was treated as any other agricultural commodity and sold in the legitimate market. To be sure, there were traders who smuggled bales of coca leaf out of the valley on the backs of mules, but only to avoid the government tax. Coca leaves have been chewed in the Andes for centuries; they are appreciated by workers for their ability to assuage hunger and to numb the pain of hard labor. Perú has always condoned a legitimate market for coca leaf, but not for refined cocaine. Cocaine is a vice of the modern world.

By the mid-1970s, the dark clouds of the drug boom had begun to gather, and one by one the friends I had made in the Apurímac were forced to flee for their lives. The entire valley was gripped in terror as innocent citizens became inextricably caught up in the struggle between government forces and increasingly powerful bands of *narcotraficantes*. It was impossible to remain neutral in this struggle. Government forces routinely threatened people at gunpoint, demanding that they serve as informers. Anyone called in for questioning by government investigators was suspected of having divulged information and risked reprisals from the other side. Anyone involved in producing, processing, or transporting *pichicata* (cocaine) viewed anyone who was not as a threat. If you weren't with them, you had to be against them. There was no in-between. Bodies lying beside the road or floating down the river served as frequent reminders that like it or not, everyone was involved.

With the rise of the Sendero Luminoso revolutionary movement in the early 1980s, the alignment of forces in the valley shifted, but the struggle continued. The revolutionaries found natural bedfellows in the *narcotraficantes*. The Senderistas cared about ideology but not about drugs, whereas the *narcos* cared about drugs but not about ideology. Complementarity of interests led to an unholy alliance in which the guerrillas tied down the police and military, confining them to heavily fortified positions at night and leaving the *narcos* free to pursue their business unhindered. In return, the *narcos* shared their profits with the Senderistas, who used the funds to finance their movement. Throughout the entire period from the late 1970s through the early 1990s, when the government of Alberto Fujimori finally restored security to Perú, the Apurímac valley, a peaceful wilderness only a short time before, was too dangerous a place for the likes of me—or, for that matter, any outsider.

Being witness to the explosive destruction of rain forest carried out in the name of development, the blatant disregard of indigenous rights, the corrupting influence of drug traffic, and the abject fear that gripped Perú during the heyday of the Sendero Luminoso has profoundly influenced my thinking about conservation. These experiences, and many others I shall relate, have convinced me of the extreme challenges ahead in the effort to conserve some bits of tropical nature for posterity. Poverty, corruption, abuse of power, political instability, and a frenzied scramble for quick riches are common denominators of the social condition of developing countries around the world. It is in this vastly different social context, contrasting in nearly every respect with the comfortable conditions we enjoy in the United States, that tropical nature must be conserved if it is to be conserved at all.

Tropical forests have been a consuming passion for me throughout my career as an academic scientist. I have spent nearly a third of my adult life in the forest, living in makeshift bush camps or at a rustic research station in Perú's Manu National Park, which I discovered in 1973 after fleeing the Apurímac valley. Located in a distant corner of the upper Amazon basin, the Manu is relatively inaccessible and so represents tropical nature largely as it existed before humans intruded into the scene.

Beyond my experiences in the Manu, I have had occasion to visit tropical forests all over the world and to view the problems of tropical conservation at first hand. These experiences have given me a global perspective on biological preservation that few people are privileged to have. What I have seen convinces me that the conventional wisdom now being applied to the conservation of tropical nature is misguided and doomed to failure.

As a tropical biologist, I have been invited to serve on the boards of directors of a number of international conservation organizations. Such organizations typically appoint a token scientist or two to provide a point of view that would otherwise be lacking. Serving on these boards has been an eye-opening experience for an ivory-tower academic who is most at home in the rain forest and the classroom. As a naive outsider ushered into the company of some of the titans of society—politicians, bankers, chief executive officers, heads of foundations, the fabulously wealthy—at first I felt out of my depth. But after sitting through a few meetings, I realized that some of the organizations were rudderless ships, lacking both vision and knowledge. Many of my fellow board members had barely stepped into the developing world, and then only to visit game parks or to dine with presidents and government ministers. Few, if any, of them had seen how a developing country looks from the bottom. Moreover, none of them, including me, was a conservation professional. Most of them specialized in financial management and the bottom line.

Board meetings were largely devoted to lengthy discussions of the financial state of the organization: what was being done to raise money; whether budget goals could be achieved; what the president's compensation should be; which public relations firm should be hired for the next campaign; whether to rent more office space or construct a new building; and endless matters of this sort. Where were the deep discussions of conservation policy and strategy? They simply were not on the agenda.

It is said that some corporate executives can't see beyond the next quarterly report, an attitude that prevails to an unfortunate degree on the boards of conservation organizations. Boards want to see results. Goals therefore cannot be too distant, and above all, they should be concrete or, better yet, quantifiable. Approaches and directions are often fad or crisis driven, reflecting the public's short attention span and the organizations' unending need to attract new members.

The world suffers no shortage of conservation crises. Elephants, rhinoceroses, and tigers are under relentless assault for their highly valued body parts. Around the world, it is easy to raise the alarm for a disappearing parrot here, a rare crane there, a vanishing tortoise somewhere else. Conservation emergencies such as these are the stuff of fund-raising campaigns, but they don't add up to a coherent plan for saving nature.

To be fair, the officers of conservation organizations are obliged to walk a tightrope between the shifting whims of a fickle public on one side and the narrow agendas of major donors, such as the billion-dollar foundations, the U.S. Agency for International Development

(USAID), and the World Bank, on the other. The freedom from financial concerns needed to enable them to sit down, chart a course, and then stick to it simply does not exist. Conservation organizations thus become prisoners of the bottom line, much as corporations are.

Faced with constantly shifting fads in conservation policy, a desire to impress directors with short-term results, an unending need to respond to crises, and an almost obsessive preoccupation with the demands of fund-raising, officers of conservation organizations are distracted from thinking deeply about ways in which conservation can be achieved over the long term. Yet no other institutions capable of crafting a global conservation strategy exist.

Seeing all this from an insider's point of view has filled me with apprehension. What can be done to ensure that nature survives the twenty-first century? That is the central question of this book, to which each chapter provides a partial answer. But from this one question come others. Is conservation being successfully implemented in the areas where most of the earth's biological wealth resides? If not, what will be required to ensure success, when success is defined as preserving the earth's wealth of wild species for the next 100 years, just to start? Can science provide adequate guidance? What types of institutions should champion the cause? How many obstacles lie in the way of our creating institutions that can sustain nature through the twenty-first century? Can the obstacles be diminished or removed? Having spent much of my adult life in tropical South America, I am perhaps more keenly aware than most people of what the challenges are. But no one, including me, has a magic wand that will make the challenges disappear. I can only offer some suggestions and hope that conservation organizations and governments will respond by designing programs robust enough to endure a century of unprecedented social and technological change.

2

ASSESSING
THE PRESENT

My eyes gaze across a pea green lake ruffled by the midday breeze under a cloud-dappled tropical sky. Shattering the quiet are the shrieks and whistles of a family of giant otters fishing somewhere up the lake. Two baby otters bleat insistently as they beg for a fish just caught by one of the adults. Across the narrow ribbon of water rises a continuous wall of virgin forest. Extending into the water, like little feet at the bases of the trees, are spring green mats of grass from which emerge the trills and cackles of foraging rails and gallinules. A cormorant suns its wings on an emergent branch, having just swallowed a *carachama,* a type of armored catfish that abounds in Amazonian lakes. To my left, a towering forest looms over my lakeside office, its edge a tapestry of vines and branches that offer thoroughfare to throngs of monkeys. Long accustomed to the benign presence of humans in their midst, they parade before my view, hardly more than an arm's reach away on the other side of the screening.

To reach this inner sanctum of nature, one must travel for three days in a motorized dugout canoe. The nearest permanent settlement is 200 kilometers away. On one side, the forest extends unbroken for 60 kilometers to the Andes; on the other, to the Brazilian border and beyond. Between these bounds lies some of the most pristine wilderness left on earth. The isolation and intimate contact with nature I enjoy here have become extreme rarities in today's overcrowded world.

For twenty-five years, I have come here to the Cocha Cashu Biological Station in Perú's Manu National Park to study tropical nature. Within the park stand groves of giant mahogany trees that astonish visiting Peruvians, to whom mahoganies are a storied relict of the past, as bison and chestnuts are to North Americans. Visiting scientists experience the rare privilege of conducting research on an intact flora and fauna. No species is known to have gone extinct here in modern times, and the forest is free of exotic (non-native) plants and animals, with only one exception: the Africanized honeybee. Honeybees are an Old World import into the Americas, and until 1978 they had not spread across this part of the Amazon. Now their presence reminds us that no place is too remote to escape the consequences of human activity.

Here, nature seems secure in a 1.5 million–hectare[1] national park, protected from external threats by a corps of thirty guards. If any tropical preserve has a chance of remaining intact and undisturbed, surely it is the Manu. But are the park and its prodigious biological diversity really secure, or do the trappings of security merely create the illusion? A distinction must be made between the park as a physical entity and the park as a human institution. The park was intended to exist forever, or so it is decreed in the formal documents of establishment, but no human institution endures forever.

Already there are signs of weakness, cracks in the dam that foretell dangers ahead. If tropical nature cannot be made secure in this far corner of the Amazon, where can it be? I cannot say, but I do know that trouble lies ahead because of a fatal disjunction in values and capacity. The well-organized societies of the industrialized world are the ones most concerned about biological diversity and most capable of providing the stable, long-term institutional support needed to preserve it. But much of the earth's biological wealth is confined to the Tropics, especially tropical forests, nearly all of which are located in developing countries where appreciation of wild nature is minimal and public institutions are notoriously frail. For many residents of the Tropics, nature has only utilitarian value, as an immediate source of wealth or a livelihood. The thought that there might be more exalted reasons for nature to exist has not entered the consciousness of many people who live in and around tropical forests.

The freewheeling societies of many tropical countries (those situated partly or entirely between the Tropics of Cancer and Capricorn) simmer with ethnic tensions and are rife with corruption and lawlessness. Their overall lack of social, economic, and political maturity results in institutional weakness and instability. Instability encourages short-term thinking, but nature protection is not a short-term proposition. If it is not forever, why bother? Preservation of tropical nature for posterity requires rock-solid institutions. Wavering political resolve and lapses of attention can be disastrous because extinction allows no second chance. There is a large measure of quixotic hubris in trusting human institutions to prevent something that is truly irrevocable. Unfortunately, there is no alternative.

Many would agree that the paramount conservation goal should be the perpetuation of biological diversity, often called biodiversity. In the public mind, the term *biodiversity* has become confused because many specialists, eager to relate their own endeavors to conservation, have broadened the concept to include diversity of genes, populations, varieties, species, and even ecosystems. My preference is to equate biodiversity with species because species are the end products of evolution and species can and do go extinct. Therefore, conserving biodiversity means establishing conditions that will serve to minimize future extinctions.

When the emphasis is placed on extinctions, the challenge of conserving biodiversity becomes more focused. Traditional approaches assume that protecting a piece of habitat inhabited by a certain species will ensure the continued existence of that species. However, this assumption is challenged by disturbing new evidence from studies being conducted in widely scattered parts of the world.

One such study was conducted in the Middlesex Fells Reservation (West), a 400-hectare (approximately 1,000-acre) suburban park on the outskirts of Boston, Massachusetts. The Middlesex Fells is one of an extremely small number of protected areas to have been thoroughly inventoried early in its history. In 1894, two of the most eminent botanists in the United States at that time, Merritt Fernald and Liberty Hyde Bailey, documented the presence of 422 plant species, including trees, shrubs, vines, herbs, and ferns.

Ninety-nine years later, in 1993, Brian Drayton and Richard Primack of Boston University resurveyed the Fells. Despite a search that covered every corner of the reserve, they failed to locate 155 of the species that had been present at the first survey, 37 percent of the 1894 list.[2] Among the missing species were wildflower favorites such as the fringed orchid and Hooker's orchid, wood lily, and cardinal flower. Partially compensating for the missing plants were 64 species that had not been recorded by Fernald and Bailey, a majority of them exotics, that is, non-native plants. It is likely that most of these, rather than being overlooked in the first survey, had invaded during the interim. Over the ninety-nine years, the proportion of native species in the flora had dwindled from 83 percent to 74 percent. The populations of many remaining native species had shrunk to one or a few tiny remnants, suggesting that further losses are in store.

Why had more than a third of the species recorded by Fernald and

Bailey disappeared in the span of a century? Drayton and Primack could offer only hypothetical explanations. Undoubtedly, the reserve had been exposed to increasing human use as it was engulfed by Boston's expanding urban perimeter. Perhaps trampling or occasional ground fires started by careless visitors were to blame. In addition, maturation of parts of the forest might have eliminated some species. But apart from such speculations, the authors could point to no single, overriding cause.

The loss of species from an enclave as small as the Middlesex Fells may seem trivial in a global sense, but the findings have ominous implications. Other studies suggest that similar losses of species are occurring in nominally protected areas elsewhere, a trend that I fear represents the tip of an iceberg. Unfortunately, no one knows the prospects for long-term survival of populations in habitat islands such as the Middlesex Fells, in part because crucial "before" data simply do not exist. Yet many of our parks are now isolated remnants of natural or seminatural habitat surrounded by urban development, agricultural land, or tree plantations. Undoubtedly, species are disappearing from hundreds of other protected areas in the United States and elsewhere, but there is little evidence to confirm this. Nevertheless, the case of the Middlesex Fells makes the point that merely including species within a protected area by no means ensures their long-term perpetuation.

Perpetuating biodiversity over the long run requires more than setting aside parks and reserves; it requires preserving intact those processes that have maintained the biodiversity of undisturbed ecosystems over past millennia. Predation, pollination, parasitism, seed dispersal, and herbivory are among these vital processes. Most of these processes are familiar ones, even to laymen. All of them involve interactions among species: animals with animals, animals with plants, plants with plants.

It is this web of interactions among species that I define as nature. Mine is an unconventional definition. Many people would define nature as the collection of plants and animals found in the outdoor world or considered aesthetically appealing in a sphere not influenced by humans. Yet such definitions lack any sense of dynamism. What one sees in a pristine landscape is not even remotely analogous to a painting, which is forever frozen in time. The living scene that greets the eye is a product of dynamic forces, the operation of which is largely hidden from view. Most likely, the scene viewed from a particular vantage point today will differ markedly from the scene that existed a thousand years ago or that will exist a thousand years in the future. Certainly, none of the individual organisms will be the same. Nevertheless, there is a continuity of process that ties the future to the past and that maintains the blend of species over time so that an oak forest remains an oak forest, even after all the individual trees die and are replaced by others. Nature as I define it is a web of interactions involving plants and animals in different combinations and in various relationships. Disrupting or distorting these interactions leads to imbalances in the functioning of the system, the inevitable result of which is species loss and simplification.

Any strategy to conserve biodiversity must maintain the web of interactions that regulates and perpetuates the ecological system. Among the most critical of these interactions are those at the top of the food chain. Predator-prey relationships, in particular, take place on very large spatial scales. A single wolf pack requires hundreds of square kilometers to maintain itself. A female jaguar needs at least 20 square kilometers to support herself and her cubs.[3] The area needed to sustain a genetically viable population of jaguars is several thousand square kilometers. Unfortunately, relatively few parks are this large, and consequently, many established parks have lost their top predators. In the absence of the ecological function performed by top predators, the whole ecosystem slides into imbalance and begins to spiral down in a cascade of species losses. Maintaining top predators, or restoring them to ecosystems from which they have been eliminated by human persecution, is but one of the challenges facing conservationists.

A great many of the world's parks and nature preserves are challenged in this way. Not large enough to maintain populations of top predators, they are unable to sustain the full spectrum of ecological functions needed to preserve biodiversity. Preventing extinctions in areas lacking the full complement of ecological processes will require ceaseless human intervention. Experience with endangered species such as the whooping crane, the Puerto Rican parrot, the California condor, and the Hawaiian crow teaches us that such interventions are extremely expensive. Thus, they are not a reasonable option for the tropical countries that contain most of the world's biodiversity.

Unfortunately, some people view the impending extinction crisis as bogus, merely a specter created by conservation organizations to further their fund-raising efforts. But can anyone be blamed for such misguided thinking? So few extinctions are publicized that the seriousness of the problem is easily underrated. Many schoolchildren in the United States could tell you, for example, that the passenger pigeon and the ivory-billed woodpecker are extinct, but few would know that roughly 20 percent of the world's bird species have been driven to extinction within the past few thousand years as humans have spread around the globe.[4] Time lags also undermine public awareness of the seriousness of today's extinction crisis. Years, even decades, may pass between the disruption of a habitat and ensuing extinctions, delays between cause and effect that represent a ticking time bomb. Large numbers of extinctions are unavoidably ahead as consequences of actions already completed in the past. So-called secondary extinctions, those resulting from habitat fragmentation and the consequent disruption of ecological processes, are particularly insidious because they are driven by processes, invisible to the layman, that operate with time lags of decades.

Arousing public concern over something that is yet to happen is difficult at best. Witness the fact that people in the United States eagerly build houses near the San Andreas Fault and on hurricane-prone coastlines along the Atlantic seaboard. One process with built-in time lags that is studiously ignored by most of the world's leaders is the growth of the human population. Overpopulation threatens the well-being, if not the outright existence, of billions of people. Yet few politicians publicly acknowledge the population problem. To the contrary, the United States Congress recently voted to cut off funds in support of international family planning programs.[5] Unless the threat is immediate and dire, like the fearsome specter of nuclear war, people predictably turn their heads to avoid confronting a difficult issue.

Indeed, the greatest challenges of conservation involve nonscientific issues: overpopulation, inequities of power and wealth, exhaustion of natural resources, corruption, lawlessness, poverty, social unrest. Building a bulwark of security around the last remnants of tropical nature requires facing these issues head-on. Doing so represents a challenge of the first magnitude, with success predicated on simultaneous progress on multiple fronts: scientific, economic, social, and political.

As a prerequisite, it is essential to get the science right, for no effort in the social realm will matter if the conditions humankind offers nature are inadequate to ensure its survival. Unfortunately, no simple prescription exists because the relevant science is still in its infancy. Conservation biology came into being only in 1978. Therefore, scientists are just now beginning to understand how extinctions can follow as unintended consequences of distortions or disruptions in the processes of nature.

Ultimately, the issue boils down to habitat—how much for humans and how much for nature. In the late 1990s, the global balance stands at roughly 5 percent for nature (counting only parks and other strict nature preserves) and 95 percent for humans.[6] Turn this ratio around and we confront the fact that 95 percent of the space once occupied by natural communities has been preempted for human use—a truly catastrophic loss for nature, guaranteed to result in countless extinctions.

Can anything be done to ameliorate the loss of habitat? Wishful thinking has produced several ideas. Many conservation writers, for example, have suggested that habitat loss need not be so great if tropical forests could jointly serve the purposes of humankind and nature. The tacit assumption is that biodiversity will no longer be threatened if humans can achieve sustainable development. A number of possibilities have been proposed, among them extraction of nontimber forest products, such as fruits, nuts, resins, and medicinal plants; regulated harvest of game and other animal products; ecotourism; and sustained management of natural forests for timber. Whether proposals to use tropical forests in these ways can become a reality or whether they represent nothing more than wishful thinking depends on a host of underlying assumptions about yields, biological sustainability, structure and availability of markets, substitutability of products, potential for mass production under cultivation, and, most fundamentally, the opportunity cost of retaining forests. Unfortunately, close scrutiny of these proposals leads to the sad conclusion that if economics rules, most tropical forests are worth more dead than alive. In the absence of sustainable development, humans will inevitably eliminate most tropical forests. But the converse notion—that sustainable development will lead inexorably to the harmonious coexistence of humankind and nature—is patent nonsense.

Despite a great deal of political lip service given to extolling sustainable development, no country on earth has achieved it or anything approaching it. Our hunter-gatherer forebears lived in a state of sustainable development, but they abandoned it as soon as they adopted modern technology and medicine. Around the world, a few countries have come close to satisfying some criteria of sustainable development while remaining far from the goal with respect to others. For example, Holland and Denmark, two of the most prosperous countries on earth, are close to sustainability with respect to population and agricultural self-sufficiency, but they are far from it with respect to consumption of energy and a wide range of natural resources. Moreover, neither country contains wildlands worthy of the name. Large animals, other than domestic stock, are a thing of the past outside of a few special game parks. Nature, in the way I define it, has been degraded to a rudimentary state in these two countries. Yet human life goes on, at a comfort level envied by much of the rest of the world. So let us recognize a fact of life: wild nature and the biodiversity it perpetuates are not a necessity for humans; they are a luxury.

Ultimately, nature and biodiversity must be conserved for their own sakes, not because they have present utilitarian value. Essentially all the utilitarian arguments for conserving biodiversity are built on fragile assumptions that crumble under close scrutiny. Instead, the fundamental arguments for conserving nature must be spiritual and aesthetic, motivated by feelings that well up from our deepest beings. What is absolute, enduring, and irreplaceable is the primordial nourishment of our psyches afforded by a quiet walk in an ancient forest or the spectacle of a thousand snow geese against a blue sky on a crisp winter day. There are no substitutes for these things, and if they cease to exist, all the money in the world will not bring them back.

Economic forces, driven by population growth and the fervent desire of people everywhere to advance their material well-being, are inexorably eliminating the world's remaining wildlands.[7] Short of radical changes in government policy in country after country, all unprotected tropical forests appear doomed to destruction within thirty to fifty years. When that time arrives, the only remaining examples of tropical nature and, consequently, most of what remains of tropical biodiversity will reside in parks. Parks therefore stand as the final bulwark of nature in the Tropics and elsewhere.

If tropical parks could be counted on to protect nature efficiently and reliably, there might be no cause for concern. However, what I have seen on visits to more than fifty parks and nature preserves in seventeen tropical countries does not instill confidence in the future. Scores of friends, colleagues, and acquaintances who have visited other tropical parks describe conditions that many of us would regard as scandalous. Large numbers of tropical parks exist only on paper and lack delimited boundaries, administrators, or staff. In some parks I have visited, barefoot guards were obliged to hunt and fish to feed themselves and their families and sometimes went without pay for months on end. Some neglected parks have been massively and illegally invaded by farmers, loggers, and miners; in the worst cases parks have effectively ceased to exist, the land having been usurped for logging, farming, or other human uses.[8]

Deficiencies in staffing and funding are not the only concerns. The social and political milieus within which parks exist are critical factors. The plight of the gorilla illustrates my point. A charismatic megavertebrate par excellence and a human relative beloved to zoo visitors and television audiences the world around, the gorilla faces a particularly uncertain future. The species *Gorilla gorilla* is parsed by taxonomists into three subspecies. One of these, the lowland gorilla, lives in the western Congo basin; another, in the eastern Congo basin; and the third, the renowned mountain gorilla, in the volcanic belt that forms the divide between the Congo basin and the plains of East Africa. The existence of three such widely separated subspecies should guarantee the security of at least one of them, or at least that is a presumption made in conservation biology textbooks. But such academic exercises tend to discount the imponderable workings of social forces. Essentially all the world's gorillas live in seven countries: Cameroon, Gabon, Congo, the Central African Republic, the Democratic Republic of Congo (formerly Zaire), Rwanda, and Uganda.

As a thought experiment, consider the long-term survival prospects of gorilla populations in these countries in light of the social and political conditions existing in the autumn of 1997, the time of this writing. Cameroon has bartered away the timber in its remaining unprotected forests to French interests, and as logging roads push into the trackless wilderness that is home to hitherto unexploited wildlife populations, the newly created access has unleashed a frenzied demand for "bush meat," in response to which the local populations of gorillas, chimpanzees, and many other animals are being rapidly liquidated for human consumption.[9] Congo and the Central African Republic are both embroiled in civil war, a situation that invites lawlessness and the dissolution of normal social restraints. The Democratic Republic of Congo has just experienced a violent overthrow of government, and its new leader appears to be heading down the same road to authoritarianism and repression that ultimately cost his predecessor his job.[10] Rwanda, having experienced a convulsion of genocide in 1994, is being run by a minority government that is distrusted by the Hutu majority. None of these five countries seems to offer a secure haven for gorillas.

Just two countries remain, Uganda and Gabon. As of this writing, Uganda is blessed with a competent and popular leader and is a paragon of stability among central and East African states,[11] but it has so few gorillas living within its borders that in isolation they would suffer inbreeding and extreme risk of accidental extinction. Thus, of the seven countries, only Gabon remains. Gabon may offer the best bet for the gorilla's survival because it harbors a large population of the apes and has been politically stable for decades.[12] However, Gabon is actively exploiting its remaining unprotected forests and, with an economy based almost exclusively on oil exports, is living on borrowed time.[13] When the country's long-reigning dictator is replaced, what kind of conservation policies will his successor implement? No one can say. The whole business of trying to conserve gorillas in the wild is a crapshoot unless Africa suddenly acquires a level of political maturity it has not displayed heretofore. I hope I am wrong, but if I had to bet, I would wager that the last gorilla will die in a zoo.

What can be done to prevent the progressive degradation of tropical parks and loss of the biodiversity they contain? This is a difficult question without simple answers, but one thing is clear: many of the approaches currently being promoted by the international conservation community fail to address fundamental issues. Conservation organizations, largely for political reasons, are obliged to view the world through rose-colored glasses. Pessimistic projections do not attract donations, and hardheaded assessments of a nation's prospects for political stability do not create goodwill with the government in power. Conservation organizations consequently dissipate much of their energy in what behavioral scientists would call displacement activity, activity that is irrelevant to the threatening situation at hand but that relieves tension for the moment.

If the conservation of tropical biodiversity is to move forward in a constructive way, the reasons why so many parks falter and fail must be addressed. None of the reasons is pleasant, and openly addressing many of them would be politically incorrect. That is the bad news.

The good news is that the situation is not hopeless. There are positive steps that can be taken, but they will require, first, that we squarely face unpleasant facts and, second, that we redesign the strategy of international conservation. These are tall orders, I realize, but they can lead us to solutions—and it is this prospect that provides the inspiration for this book.

CHAPTER 3

PARADISE FADING

To READERS accustomed to the well-kept appearance of national parks in the United States, replete with highly educated rangers who greet visitors, my concern about the condition of parks in many other countries may seem strangely alarmist. But the fact is that the National Park Service, a division of the U.S. Department of the Interior, is perhaps the best-funded and best-staffed such organization in the world. No other park service in my experience comes close to matching the U.S. standard. To illustrate the point, I shall recount some of my own experiences as a seasonal visitor to Perú's Manu National Park over the past twenty-five years.

As a repository of biodiversity, Manu National Park stands without peer. Its location on the western fringe of the Amazon basin puts it at the world's biodiversity epicenter. The park's biological value is further enhanced by its design, encompassing the entire watershed of the Manu River and its tributaries, from the 4,000-meter-high crest of the eastern Andes far out onto the lowland plain. By spanning such a broad range of environmental conditions, the Manu earns the distinction of holding more biodiversity than any other park in the world.

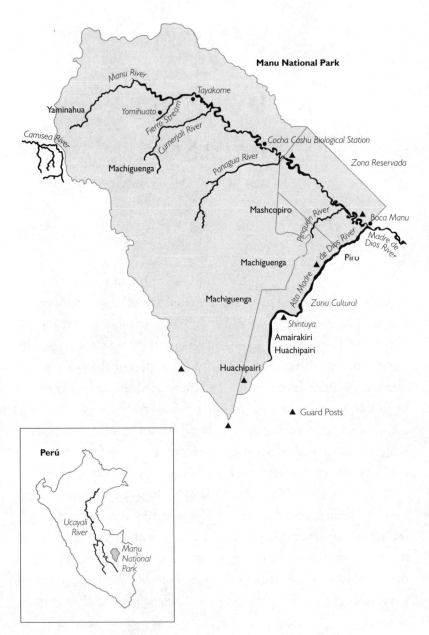

Manu Biosphere Reserve and its indigenous peoples.

Leading a litany of superlatives is a steadily expanding bird list of almost 1,000 species.[1] (By contrast, only 650 species reside in all of North America north of Mexico.) In addition, more than 200 species of mammals, including 13 primate species, jaguars, pumas, ocelots, tapirs, capybaras, giant anteaters, and spectacled bears, live within the park's boundaries.[2] Reptiles and amphibians provide another showcase of diversity. Every year, the list of species known to occur in the park notches upward. The park's lowlands can claim nearly 90 species of frogs and toads, a number surpassed only at one locality in Ecuador.[3] Tree diversity in the Manu's forests ranges from 150 to 200 species per hectare.[4] In just one month, a team of expert lepidopterists documented more than 1,300 butterfly species at a single lowland site.[5] I could undoubtedly go on and on with such boasts had other groups of organisms been so thoroughly inventoried.[6]

Such unsurpassed biological richness has brought me back to Manu National Park year after year. For two and a half decades, I have managed a small biological research station in the core of the park, la Estación Biológica de Cocha Cashu (Cocha Cashu Biological Station). I reside there for three months each year, usually from September through November. The station itself is a collection of three thatch-roofed buildings, all of them constructed from locally available materials. Two small banks of solar panels provide energy for computers, fans, electric lights, and radio communications. Having no institutional support, either from within Perú or from foreign sources, the station is perforce financially self-sustaining. Fees levied on investigators generate the revenues that provide for one full-time employee and two part-time employees. Limited resources dictate a policy of minimalism. In one respect, that is good. Too much money would inevitably lead to diminished isolation and spoil the rustic tranquility that is one of the delights of being there.

Notwithstanding its physical remoteness and limited facilities, the station has built a gratifying record of scientific productivity, thanks largely to the industry of more than two dozen graduate students who have completed doctoral research projects at the site. The station has also played host to more than 100 Peruvian university students, many of whom have carried out thesis projects for bachelor's or *licenciatura* (similar to master's) degrees.

From my vantage point at the research station, I have had the opportunity to observe firsthand the workings of a national park in a developing tropical country. Although scientists have been bystanders rather than participants in the park's development, the station's location on the main trunk of the Manu River and the year-round activity it sustains make it a focal point in the region. Boats going upriver or downriver for whatever reason usually stop, and the passersby brief us on local happenings. The station is socially integrated into the park community, which includes ecotourism operators, park guards, administrators, and members of several indigenous communities.

Because of its extraordinary biodiversity, the Manu enjoys an almost mythical reputation and has been accorded the highest rank among Perú's national parks in the competition for scarce budgetary resources. Since its beginning, the Manu has benefited from a corps of thirty guards, who, at great cost to their personal lives, have maintained an official presence at remote outposts along the roadless boundary. Burdened with numerous handicaps, often poorly educated, and always woefully underpaid, the guards have nevertheless admirably performed their duty of keeping the outside world at bay.

Yet in my view, the park is under dire threat of extinction as a haven for nature. The park is a victim of best intentions gone awry, of a shortsighted administration, of inconsistent policies, and now of a situation slowly spiraling out of control. Because the problems of the Manu are ones that are likely to reappear elsewhere in the Tropics, they warrant close scrutiny. At issue is a moral dilemma that pits the rights of our own species against the survival of nature. It is a contest that nature nearly always loses.

To understand why the Manu is approaching a critical juncture requires historical perspective. The Manu River, which is the gateway to the park, was unknown to Europeans until the 1890s, when the burgeoning rubber boom brought ruthless opportunists and fortune seekers into the region to prospect for forests rich in rubber trees. It can be presumed that the Manu basin then supported a sizable indigenous population, as did the basins of other Amazonian white-water rivers before Western contact, but the rubber barons left few written records.

For several decades, the Manu River served as a major thoroughfare in the rubber trade, a crucial link between the rubber-rich forests in the eastern Madre de Dios River basin and adjacent Bolivia and the markets of Iquitos, where the product was loaded aboard oceangoing vessels for shipment to Europe and North America. The headwaters of the Río Manu represented the Northwest Passage of the rubber trade, a circumstance brought about by a geographically rare juxtaposition of the headwaters of two major river systems. A neck of land only two kilometers wide at its narrowest forms the divide between waters flowing east into the Manu and west into the Camisea River, a tributary of the Ucayali, Amazonian Perú's main axis and a river that rivals the Mississippi in its breadth and flow.

Shortly after the rubber boom collapsed, around 1912, the Manu became depopulated. A few outsiders lingered into the 1940s, but the indigenous populations that had occupied the region before the boom had vanished. Many of the Manu's indigenous people had been pressed into the service of egomaniacal rubber barons, as so vividly portrayed by Werner Herzog in his 1982 film *Fitzcarraldo*. Many others undoubtedly died from diseases introduced by the rubber traders, as happened nearly everywhere on the first contact of Europeans with indigenous peoples.[7] The lucky few who escaped disease and enslavement fled to take up a fugitive existence in remote headwater tributaries inaccessible even to small dugout canoes. For a hundred years, these refugees from the rubber boom and their descendants have lived in total isolation from the outside world.

I was once privileged to glimpse the fugitive lives of the park's un-contacted human residents. My companions and I had just completed a two-day trip in the back of an open truck. We were on our way to the Cocha Cashu Biological Station with a load of supplies and were spending the night at a tiny Dominican mission where the road ended, a place called Shintuya. In the morning, we were to be met by the sta-tion's "canoe," a sixteen-meter (fifty-two-foot) dugout built up with sideboards to increase its draft. Capable of carrying several tons of food, fuel, equipment, and personnel, these river craft are the eighteen-wheelers of the Amazon. From the mission, we would travel another three days on the river before we reached the research station.

By chance, our arrival fell on July 27, the eve of Perú's indepen-dence day, so everyone was in a festive mood. The ramshackle, dirt-floored cabin that served as the local tavern was packed with truck drivers and local residents reveling in boisterous conversation as the tables gradually filled with empty beer bottles.

I try to avoid such gatherings because once one joins a drinking group, local etiquette calls for the downing of bottle after bottle until one is in a stupor. That night, every clique of drinkers was insisting that I sit down and socialize. Had I complied with the invitations, I would have wound up in a horizontal condition, something I wanted to avoid at all cost. So I abandoned the festivities and went out into the starry night.

The only other person out there in the dark was a Machiguenga, a member of one of the local ethnic groups. I don't remember how our conversation started, but both of us had consumed enough beer to be in a garrulous mood, so the talk flowed freely. He had been born at the headwaters of the Río Pinquén, a Manu tributary, he recounted. He lived in an extended family group, the traditional unit of Machiguenga social organization. No one in his group had ever met an outsider, and they all lived in perpetual fear of blundering into a white person by accident. After three generations, the terror of the rubber boom was still indelibly imprinted on their collective psyche.

When group members traveled on hunting trips, they made a practice of sleeping in a different spot every night. Each morning, the remains of the previous night's campfire were carefully dispersed and covered with dead leaves. When they crossed a stream, they erased their footprints. Such precautions were taken with the conviction that anyone they might happen to meet would shoot them on sight. I don't recall how my acquaintance said his group had been contacted, but it had happened when he was still a boy. He later attended a mission school, where he learned to converse in Spanish. By the time I met him, he had been assimilated into the local population, one of millions of Peruvians of Amerindian descent.

Until the 1960s, the land along the entire main channel of the Río Manu, including its navigable tributaries, was unpopulated. No indigenous group wanted to live beside a navigable river because a boat carrying hostile strangers could appear at any time without warning. The barrier of fear that isolated the Manu's indigenous people began to crumble, however, with the arrival of a Texas-based international missionary group known as the Summer Institute of Linguistics (SIL), which is associated with the Wycliffe Bible Translators. The SIL established a post on the Río Manu at a location the missionaries named Tayakome, near the geographic center of what is now Manu National Park. They brought with them Machiguengas from another region to serve as intermediaries in establishing contact with Machiguenga groups at the Manu's headwaters.

Wycliffe missionaries have proselytized indigenous groups all over the world, driven by the conviction that the Second Coming will occur when all the world's peoples have been introduced to the Bible. Hoping to hasten that day, they aim to carry the Bible to every ethnic group on earth. In pursuit of this goal, Wycliffe's sister organization, the SIL, has become renowned for its ability to crack hitherto unknown languages. When a new group is contacted, SIL missionaries are sent in to work with the people, and within a few weeks they decipher their language. Next, a dictionary is compiled, and eventually the Bible is translated into the hitherto unwritten tongue. The people are then taught literacy in their own language, with the primary aim of having them read the Bible. The program may suit the designs of the SIL mission, but it results in disillusionment for the villagers when, with increasing awareness of the outside world, they discover that they have acquired literacy in a language no one else speaks, reads, or writes.

Many missionary organizations are chronically underfinanced. Missionaries view their work as a calling and often endure privation for the sake of realizing their nonmaterial ambitions. Although the SIL headquarters at Yarina Cocha in central Perú has all the trappings of a middle-class U.S. suburb—in startling contrast to the squalor of the neighboring Peruvian town of Pucallpa—the imperative of the budget, at least in the 1960s, compelled a need for entrepreneurial activity.

One way SIL missionaries at Tayakome generated income was through the sale of pelts and hides. The men of the village were issued shotguns but only a limited number of cartridges. A shotgun compared with the bows and arrows that constituted the villagers' traditional arsenal is like a modern guided missile next to an old-fashioned muzzle loader. To the Indians, firearms represented an exhilarating quantum leap in technology. Monkeys that had heretofore been safe in the top branches of tall trees and peccaries that could flee when impaled with an arrow were now certain prey. Meat could be had every day in abundance, and for many forest-dwelling people, meat is the centerpiece of the diet, relished in whatever quantity it can be obtained. But the shotguns came with a hitch. Replacement cartridges could be obtained only in trade for skins and hides of peccaries, jaguars, ocelots, giant otters, crocodilians, and other species, so hunters became entrapped in a perpetual cycle of dependence. The pelts they traded for cartridges were then sold to help finance the missionaries.

While SIL missionaries were expanding their influence among the widely scattered indigenous groups of the Manu and its tributaries, a plan was being promoted in faraway Lima to make the entire area a national park. By 1970, the plan had acquired the status of a project within the Peruvian government, allowing it to receive external funds. The project was being financed by the World Wide Fund for Nature and other international sources.

At the time of the park's formal inauguration, in August 1973, three former professors at Perú's elite National Agrarian University were in key positions in the Ministry of Agriculture's General Directorate of Forestry and Wildlife (Dirección General Forestal y de Fauna, or DGFF), the division of government then responsible for national parks. The halcyon days of the park's early period are a credit to the inspired leadership of these individuals, Marc Duorojeanni, Carlos Ponce, and Antonio Brack, backed by their academic colleague Manuel Rios, who wrote the park's management plan.

Even before the date on which the park became a formal reality, the government had taken steps to isolate Tayakome by felling trees onto its airstrip. The purpose was to prevent the missionaries from smuggling shotguns and ammunition to the Machiguengas. A rustic *puesto de vigilancia* (guard post) was built about a kilometer away and staffed with two guards, expressly to ensure that the airstrip was not reopened. Today, tall trees occupy the site of the former airstrip, and to my knowledge, all the Machiguengas living within the park now hunt only with bows and arrows.

The energy and conviction of the park's first administrators were decisive in ridding the park of another threat. Three sawmills were operating on the lower Manu River when the park was inaugurated. The presence of these sawmills influenced the drawing of the park's boundary because it was believed that the core park should be pristine. At the time, the lower Manu had been exploited for *cedro,* a type of mahogany, upriver as far as the mouth of a tributary, the Río Panagua. The park boundary was thus set accordingly, to include only the unlogged upper Manu basin. The area downstream, a large tract amounting to 257,000 hectares, was consigned to legal limbo as a *zona reservada* (reserved zone).

The park's savvy administrators took advantage of their control over the *zona reservada* to administer it as an integral part of the park, applying the same laws, controls, and restrictions. Exercising such hubris allowed them to shut down the sawmills and, effectively, to expand the park. The sawmill operators were given a one-year grace period and told to withdraw all their equipment and personnel by the year's end. Force of conviction proved its worth: as an eyewitness, I can affirm that the sawmills were indeed gone when the year expired. Now, twenty-four years later, *cedros* again line the banks of the lower Manu. Signs of the earlier logging are invisible from the river, and the only evidence in the forest are scattered rotting stumps in an otherwise unaltered forest.

Manu National Park was launched with international fanfare befitting its premier size and biological importance among the world's national parks. Financial assistance poured in from the World Wide Fund for Nature, enabling the Ministry of Agriculture to implement the park in a way that could not have been imagined with the budget appropriated by the government. *Puestos de vigilancia* were constructed at strategic access points, a corps of thirty guards was hired and trained, and boats, motors, and vehicles were purchased. The Manu was not just a paper park—a park that exists only in government documents—like so many others in the developing world; it was a real park with an administration, staff, and infrastructure. The guards were not highly educated, but they were inspired and willing. The future looked bright.

In Perú, as in all countries, political environments evolve. Within a few years, the strong support the administrators of the DGFF had drawn from higher levels began to weaken, a victim of frequent cabinet reorganizations and personnel shuffles at upper ranks. Discouraged, the original triumvirate of administrators resigned, one by one, to take more lucrative and less frustrating jobs elsewhere. Without the oversight of this committed group of advocates, conditions in the park began a long downhill slide. The local director, a man of legendary indecisiveness and inactivity, created a vacuum of leadership.

The lack of leadership in turn had a direct effect on the very people responsible for the park's well-being, that is, the stalwart park guards. Until then, the guards had experienced predictable, if spartan, conditions. Once a month, a supply boat would arrive at their far-flung outposts, bringing food, pay, fuel, and other essentials. The park had no radios, but a functioning boat was on hand at each post in case of emergency. Guards served two-month rotations, and between rotations they could look forward to a month with their families, most of whom lived several days' travel away in major towns. It wasn't a life to be envied, but the guards saw themselves as pioneers in a new venture and were inspired by the pep talks they received several times a year when the big bosses from the Ministry of Agriculture came around on inspection tours. All this gradually changed.

With new administrators filling key positions in the ministry, visits to the park by high-level officials became a rarity. Lack of oversight and a dearth of leadership contributed to a growing sense of apathy among the guards. They complained that no one seemed to care about them anymore, and they were right. Whereas previously they could count on scheduled supply boats and regular home leave, unpredictability now reigned and privation became routine. Supply boats failed to appear on schedule. With no radio communication, the guards were at a loss to know why, or when one could be expected. Sometimes, three or four months would go by without word from headquarters. With food supplies exhausted, the guards were obliged to hunt and fish to feed themselves. Many times, I arrived at Pakitza, the guard post nearest the research station, to find one guard on hand while the other fished in the river so they could eat that night. Under such circumstances, self-preservation understandably took priority over the needs of the park.

In the hierarchical culture of Perú, no one as lowly as a park guard dared complain. Protesting one's conditions to higher administrative levels would have been unthinkably brazen, almost certain to result in loss of employment or worse. Mostly the guards endured in silence, except on rare visits home, where all too often they met with unexpected family crises. Some resigned, but a good many carried on, unable to find alternative employment in an economy debilitated by hyperinflation and the antibusiness policies of a socialist government.

Despite a weak administration and lukewarm support from the central government, the park's existence was never in doubt during the early years because the military regime that ruled Perú from 1968 to 1980 offered a measure of stability and continuity of policy. During that time, however, the government instituted a radical set of socialist policies that destroyed the country's agricultural and industrial sectors and, in the end, bankrupted the economy. Fiscal irresponsibility led to hyperinflation and economic stagnation. Toward the end of the military period, relentlessly rising prices and frozen wages made the regime enormously unpopular with the public. I well remember talking with Lima taxi drivers whose animosity toward *los generales* was so passionate that some could only sputter incoherent invective and obscenities when asked about the political situation. No one wants to be that unpopular, not even Latin American generals. Finally, in 1980, after twelve years, the military command held elections that restored a civilian government.

One of the legacies handed to the civilian government by the military was a program of regionalization. A reaction to long-standing neglect of the country's rugged interior, regionalization was instituted to offset the hegemony of the capital, Lima, where roughly one-third of the population lives. Whereas the departments and provinces (counterparts of states and counties in the United States) had previously been administered by officials appointed by the central government, under regionalization the departments gained a measure of local autonomy, a condition that persists in modified form today.

Regionalization had the most profound and, in my view, negative effect on Manu National Park. Although the title of national park was retained, in effect the Manu became a regional park. In place of better-educated and more worldly appointees from Lima, the officials who presided over it—and over the park's budget—derived from the region and not the central government. Regional officials tend to be lifelong residents of the provinces, with all that implies for outlook, education, and general sophistication.

Under regionalization, every park in Perú suffered. The once influential conservation directorate in the Ministry of Agriculture was downgraded to an office of three bureaucrats whose main responsibility was to issue authorizations to exporters of wild plants and animals. They had minimal authority over the budgets and day-to-day operations of parks and other nominally protected areas. In effect, there was no national park service in Perú.

Within the Región Inka, where the Manu resides both physically and administratively, the park had few champions. The overriding preoccupation of regional officials was economic development, and many viewed the park as an unproductive lien on the region's finances. More than once, the park's budget came precariously close to being eliminated by unsympathetic regional administrators.

The worst came at the end of 1991, when Perú hit bottom. The economy was in the doldrums from two decades of ruinous mismanagement. Terrorism and a cholera outbreak early in the year had combined to bring the vital tourism industry to a standstill. As my students and I were leaving the park in November of that year, we stopped to speak with the guards at the Pakitza post. They complained that they had not received any pay in six months. Shortly afterward, I learned that the entire guard corps had been laid off for lack of funds. For the next half year, the posts were empty. By April 1992, the park had been assigned a budget by the Región Inka, and within a month a new corps of guards had been hired and installed at the posts. These untrained newcomers had none of the know-how of the experienced guards they replaced, but neither had they become jaded and cynical about the lot of a *guardaparque.*

Their lack of cynicism would not persist, however. The following November, when we were again making our annual departure, the newly hired guards complained that they had been deceived. After six months at their posts, they had yet to receive their first pay. Some talked of resigning, and later they did. More innocents were contracted to fill the vacancies. With such high turnover in the guard corps, there was little point in investing in training, even if funds for the purpose had been available.

On one stopover at Pakitza, some companions and I spoke with a park guard who was unusually forthright. After twenty minutes of small talk devoted to feeling out our views on sensitive issues, he began to unload his complaints. He was proud to be a park guard, he

declared, and he believed in the mission of the park, but the work offered no incentives. It did not matter whether or not a guard did his
job well. A conscientious guard received the same pay, status, and ad-
vancement as one who did little or nothing.

Moreover, the administration did not provide the support promised. Each post was supposed to receive fifty gallons of gasoline per
month, but in the past month, none had been delivered. In the current
month, the director himself had visited the post (a rare event) but had
left behind only forty gallons, saying that he had to deduct ten gallons
from each post to supply the needs of his journey. Consequently, the
gasoline actually received over a two-month period was only 40 percent of the nominal quota.

Of even greater concern was the food ration. As what amounted to
hardship pay, each guard was supposed to receive a monthly food
allotment valued at $82. What the guards actually received was ten
kilograms of rice, ten kilograms of sugar, and five kilograms of beans,
a ration that we later estimated would cost no more than $23 in the
retail market. Naturally, we speculated about the missing balance,
which, for twenty guards, would amount to almost $1,200 per month.
This was a tidy sum in Perú, where in 1996 the per capita gross domestic product was $1,090.[8]

Then there was the matter of uniforms. A donor had supplied
funds to the park for the purchase of uniforms. But when the uniforms arrived, they proved to be of inferior material and workman-
ship. Worse, the sizes did not correspond to those of the guards. One
pair of trousers was described as being so big that the guard to whom
they had been issued could have buckled them under his armpits. Our
informant was especially indignant about this incident, saying that it
was demeaning and indicative of an attitude of contempt toward the
guards. He confided his resolve to quit his post if conditions did not
improve by the end of the year.

Fortunately, conditions have improved markedly since the dark days of the early 1990s. Perú's economy has been on the upswing, thanks to enlightened macroeconomic policies instituted by the Fujimori government. Inflation has slowed so much that the exchange rate with the dollar has barely wavered in two years. The two guerrilla movements that kept the country under a pall of terror for a decade have been decapitated and are in disarray. Responding to these favorable developments, Perú's economy has rebounded; in 1994, it grew by 12 percent, the highest rate in Latin America. The country's improved security and financial health have inspired the confidence of international investors and donors.

Generally improving conditions in Perú had produced two positive developments for the park as of late 1997. First, prolonged negotiations had resulted in an accord whereby full responsibility for the park's administration was to pass from the Región Inka to INRENA (Instituto Nacional de Recursos Naturales), the newly constituted natural resources agency of the Fujimori government, on January 1, 1998. Second, the park's competent and committed new director, Ada Castillo Ordenola, happily announced that the World Bank's Global Environment Facility had guaranteed funds for the park's budget for the next ten years.

Under these conditions of renewal, the park's guards are visibly energized. They have new uniforms that fit. Their pay has been raised, and it is delivered on time, along with the allotted food ration. The guard posts once again have functioning outboard motors and gasoline to fuel them. Two-way radios keep each post informed of park business on a daily basis. Not since the mid-1970s have things been so good.

With such positive changes in place, I wish I could end the story here, confident that the future will take care of itself. To be sure, improved financial health is a measure of progress, but it is only a superficial one. There is better morale among the guards, whose days are occupied with the control of tourists and scientists, groups that tend to be law-abiding and solicitous of the park's well-being. Yet lying behind a facade of normalcy is the fact that the guards are hamstrung by a lack of authority to make arrests. Neither the guards nor their administrators have the power to oppose any serious challenge to the park's integrity. Consequently, the most serious threats to the park are not addressed; they are merely ignored.

CHAPTER *4*

THE DANGER
WITHIN

BUFFERED BY ITS remoteness and difficult access, Manu National Park has been relatively free of the external threats that have degraded other tropical parks—poaching, mining, invasion by squatters, illegal pasturing of livestock, and timber extraction. All these threats are present, but they are small in scale and affect only limited portions of the boundary zone. Instead, the Manu faces a much larger threat, one emanating from within its own borders: human population growth. Living within the park are communities both illegal and legal, belonging to five indigenous ethnic groups.

About sixty *colono* families occupy the park's southwestern sector. Some of them have been in the park since its inception; others have moved in subsequently. No action has been taken to remove them. The former director of the park grumbled about their presence from time to time but said that he lacked the means to evict them. Park guards with whom I have spoken expressed fear that the squatters would not acquiesce to eviction. One surmised that some of the squatters might be involved in the cocaine trade and, if so, could be well armed and willing to resist. Rather than force a confrontation, the park's administration has decided to play it safe and not press the issue. The illegal presence of sixty families in so large a park is not a great threat in itself, but it sets an unfortunate precedent.

The Manu, like many other tropical parks, also harbors a population of indigenous people. Some countries, including the United States, exclude human residents from parks as a matter of policy. Other countries, such as India, allow established villages to continue, albeit with restrictions, after an area has been decreed a park. Because forcibly removing people from parks is viewed as inhumane, the whole topic of people in parks has become highly controversial. From the standpoint of an administrator, the easiest course is to take no action at all and to proclaim that a park's residents pose no threat because they have been there for decades or centuries without degrading the area's natural resources. Is this a valid argument, or is it simply a cop-out? A twenty-five-year perspective on Manu National Park may help answer the question.

When the park was legally established in 1973, it contained several groups of contacted indigenous people and an unknown number of uncontacted people. At the time, members of the contacted communities were still living in traditional fashion, using bows and arrows to hunt and growing and spinning cotton, from which they wove cloth. They had acquired a few axes, machetes, and metal pots but few other manufactured goods. Outsiders' visits to their villages were infrequent and of brief duration.

It was freely acknowledged by the triumvirate of administrators who oversaw the park in its early days that the presence of humans represented a dire threat, perhaps not at present but certainly for the future. The issue was quite openly discussed within the conservation directorate, although to my knowledge, a concrete plan of action was never presented for formal approval. However, on my first trip to the park, in 1973, I met an anthropologist, Marcel d'Ans, a professor at

San Marcos University in Lima, who had been contracted by the park agency to open communication with uncontacted indigenous groups as a prelude to luring them out of the park.

Fortunately, d'Ans and the park guards who worked with him never came within shouting distance of an uncontacted Indian. Had they done so, they might not have lived to tell the tale. To this day, twenty-four years later, no outsider has ever initiated contact with any of the park's fugitive indigenous groups. To do so would be extremely dangerous.

We had known since our first visit to the park that uncontacted people were living in the upper reaches of the Manu. The whole watershed upstream of Tayakome was off-limits to everyone, park personnel included. It was rumored that the region was occupied by Amahuacas, a reputedly aggressive tribe, a few groups of which were in contact with missionaries near the Brazilian border. No one connected with the park had firsthand knowledge of the people who lived in the Manu headwaters. All that was known about them was that their arrows were made of bamboo and were sharp as razors.

One afternoon in the late 1970s, the two park guards stationed at Tayakome arrived at Cocha Cashu unexpectedly and in a state of extreme agitation. The Amahuacas had attacked Tayakome, they said; terrified for their lives, the guards had fled in their canoe, not stopping until they reached the research station. The next morning, they continued to Pakitza and on downstream, for in those times there were no radios in the park, either at the research station or in the *puestos de vigilancia*.

The flight of the park guards left us, a handful of unarmed researchers, feeling exposed and vulnerable. Not that the two guards would have been a match for a war party of Amahuacas had their post been attacked, but what if the marauding band continued downstream and discovered the station? We didn't want to think about it.

Fortunately, nothing happened. A few weeks later, the *guardaparques* returned to their post. They later reported that the raid of the Amahuacas had been brief and of no lasting consequence. The Machiguengas at Tayakome had avoided casualties by fleeing into the forest. Apparently the Amahuacas had been intent only on raiding the gardens of the Machiguengas, and after they had done so, they vanished.

For several years afterward, there were no further reports of interactions with uncontacted natives. Then, one day in the early 1980s, I happened to be on the riverbank when a makeshift raft carrying several people floated into view. One of the travelers noticed me and cried out for assistance. The group included a Venezuelan and some Europeans. They were adventurers, pure and simple, who had decided to retrace the route of nineteenth-century rubber baron and explorer Carlos Fitzgerald (Fitzcarrald) by making their way up the Río Camisea, over the isthmus, and down the Río Manu. Unable to duplicate Fitzcarrald's feat of hauling a boat over the isthmus, they had built a raft of balsa logs at the Manu headwaters and launched it into the river for the slow journey downstream.

The second day on the Manu, they were attacked without warning. To save themselves, they jumped off the raft and clung to the side opposite the source of the arrows. Some distance downstream, thinking they had survived the onslaught, they climbed back onto the raft, only to be attacked again. The Indians were running through the forest across loops of the meandering river to catch up with the raft, which had to run the gauntlet of repeated fusillades. By the end of the ordeal, one member of the party had been shot. Several days later, when the group reached the research station, he was bearing an ugly, suppurating wound. We treated the victim with antibiotics and helped the adventurers reach the guard post downstream.

That was the last we heard of the Amahuacas for several years, until one day in 1985. One of my graduate students, Carol Mitchell, was alone at the research station that day, her companions being off in the forest. Looking up from her work, she was riveted in surprise to see eleven naked men filing into the small clearing that contains the station's buildings. These were people she had never seen before. All the Machiguengas upstream wore clothing or *cushmas,* traditional sleeveless garments. These people wore nothing but strings of beads that ran under their noses and over their ears. Frozen in anxiety, she could only stare as they approached.

People's reactions at such moments of high tension can be strangely incongruent. Strung along the edge of the clearing was a clothesline, which was hung with drying wash. Noticing the clothing, the strangers rushed over to help themselves, grabbing piece after piece off the line. Seeing her fresh wash being stolen in front of her eyes, Carol let out a screech, rushed from the building, and snatched the clothing away from the open-mouthed intruders. Gathering it up in a bundle under her arm, she ran back into the house and slammed the door.

At this point, neither side knew what to do or what to expect. There passed an interminable moment when Carol and the Amahuacas stared at each other through the screening that substitutes for walls in the tropical climate. Flustered but having no sense of immediate danger, Carol impulsively decided that hospitality was the best policy. She came out of the main building and motioned the men to accompany her to the kitchen. There she prepared *chicha morada,* a sweet drink akin to Kool-Aid, and mimed that the visitors should help themselves. Obligingly, they passed around the pitcher of *chicha.* After they had drunk their fill, they made a ceremonious departure, embellished with unintelligible speeches in their language, and continued their journey downstream.

By the time of this incident, the station had acquired a radio, so when the rest of us returned, we alerted people downstream to expect visitors. The news was greeted with extreme apprehension by everyone in the region except for Padre Ignacio, the Basque priest who headed the Dominican mission at Shintuya. For years, Ignacio had been wanting to make first contact with a group of wild Indians. This was the moment he had so long anticipated. The next day, he was in his canoe, headed to Boca Manu, the *colono* village at the mouth of the Manu, to intercept the uncontacted Indians. At this point, everyone assumed that they were Amahuacas, but that proved to be wrong.

Ignacio ferried the group up the Río Alto Madre de Dios to the mission. His first project was to make recordings of their conversation to play over the radio to other priests and missionaries in the hope of learning what type of language they spoke. His initial attempts were unsuccessful, but after the local missionary network spread the word, a retired priest who had spent his life in the Peruvian Amazon was brought in to listen to the tapes. He recognized bits and pieces of the dialog, enough to identify it as a Panoan language. On the basis of this diagnosis, it was concluded that the unknown Indians belonged to the Yaminahua tribe. The Machiguengas speak an Arawakan language; other groups in the area speak Herakimbut, also an Arawakan language. Linguistically, southeastern Perú is one of the most complex parts of the Amazon region.

After their long journey, the Yaminahuas were hungry. Ignacio fed them amply, but their taste ran more to meat than to rice and beans. The first day, they started on Ignacio's chickens. Then it was his ducks. When they killed a cow, Ignacio decided it was time to get them out of there.

Taking them home was no simple task, even with an outboard motor to power the canoe. It is 90 kilometers from Shintuya to Boca Manu and another 200 from there to Cocha Cashu, as the river goes. On his return, Ignacio's driver told us that it had taken an additional thirty hours' journey beyond Cocha Cashu to reach the Yaminahuas' camp in the uppermost navigable portion of the Manu. The settlement, where the women and children had been left, was reported to be rudimentary compared with those of the Machiguengas. Altogether, the group numbered about thirty individuals.

Despite the loss of his livestock, Ignacio was generous with the Yaminahuas, passing out clothing, machetes, and other basic goods while trying unsuccessfully to learn what had prompted their venture into the outer world. Their Panoan language proved to be impenetrable, unknown to anyone in the vicinity. For their part, the 'Nahuas, as we came to call them, excelled as mimics and began to pick up Spanish words almost immediately.

First contact with indigenous peoples has been described as obeying the Heisenberg uncertainty principle. Contact with modern civilization irrevocably changes people, so it becomes impossible to observe them as they were before. A man who is given a steel ax will never again pick up a stone ax in the manner of his forefathers. Acquisition of one labor-saving device creates a desire for more. The gift of a mosquito net or a set of clothing creates dependence in a society that was independent until the minute before.

Contact alters life for both parties. One side is instilled with desires it never had before and a feeling of impotence and inferiority in the presence of a technology it cannot understand or acquire. The other side is pressured into a form of charity that it knows will only foster dependence. First contact is thus a no-win situation.

The station received regular visits from the 'Nahuas for several years after that startling day in 1985. With each successive visit, they spoke a few more words of Spanish, but it was never enough for a real conversation, only enough to ask for things. Unlike our neighbors the Machiguengas, with whom we have been trading for more than twenty years, the 'Nahuas had nothing to offer in return. We filled some of their requests but always reluctantly, knowing that too much generosity on our part would foster a beggar mentality in the 'Nahuas that would serve neither them nor us in the long run.

Eager to acquire new things and handicapped by their isolation in the remote headwaters of the Río Manu, the 'Nahuas have recently moved out of the park to settle at a Dominican mission on the Río Camisea, to which Padre Ignacio has been reassigned. Ignacio is teaching them how to farm and to work for what they receive. It is going to be a long, slow cultural transition that will not be accomplished without painful awakenings. Yet the 'Nahuas are luckier than they realize. A hundred years ago, the rubber barons gave no charity.

The story of the 'Nahuas is merely a footnote to the main theme of indigenous people in the park. There were only thirty contacted 'Nahuas, but there are hundreds of Machiguengas.

I have visited the main Machiguenga village of Tayakome only once, even though it is located just six hours' travel upstream from the research station. My single visit was at the invitation of the park's director, a necessary condition because the village was off-limits to anyone but park personnel, a measure intended to shield the residents from exploitation by outsiders. At the time of my visit, Tayakome had approximately 130 inhabitants, including Mauro, the schoolteacher. Mauro, himself a Machiguenga, had been educated at a mission school. At Tayakome, he received a salary and logistical support from Padre Ignacio. When I arrived, I was surprised to find that the children could speak passable Spanish, even though their parents could muster only disjointed words and phrases.

Shortly before my visit, there had been an upheaval at Tayakome, sparked by the presence of park guards only a kilometer from the village. The guards had not been taught how to conduct themselves, and they appeared to the Machiguengas as arrogant and domineering. Rumors circulated widely that guards had taken advantage of Machiguenga women. Tensions had reached the boiling point, and the park's administrator decided to defuse the situation by decommissioning the guard post. The residents of Tayakome have been left to themselves ever since.

Prior to the withdrawal of the guards, tensions within the native community led to a diaspora of its members. The pivotal issue was the amount of contact desired with the outside world. Individuals made their own choices. One large contingent, amounting to about half the families of Tayakome, chose to distance itself from Western influences by retreating up a narrow tributary, Quebrada Fierro, to form the community of Yomihuato. Yomihuato cannot be reached by motorized craft except at times of exceptionally high water, a matter of a few days per year. At other times, the journey requires hauling a dugout up rapids for two or three days. Those who chose to move to Yomihuato wanted to be beyond the reach of park authorities, and they succeeded. For nearly a decade, they remained almost completely isolated from the rest of the world.

A second contingent of families elected to remain at Tayakome, the main attraction being Mauro and his school. Even though most of the adults did not speak Spanish and had never seen an automobile or an electric light, they knew they were surrounded by an alien world, and they wanted their children to be better prepared for it than they were. In their view, Tayakome was no longer undesirable once the guard post had been decommissioned. Situated on the main river, the village is accessible by motor-powered craft in all weather. Occasional visits by park authorities provided residents with a window to the outside world and a meager supply of basic goods such as salt, matches, machetes, and clothing. In case of an epidemic or other emergency, the park could send a doctor. In contrast, the people at Yomihuato could do nothing for their sick beyond calling for the witch doctor and his herbs.

The smallest contingent of the diaspora, a mere three families, made up the progressive wing of Tayakome society. Aguilar, Zacarías, and Sandoval moved with their families to the mouth of the Río Cumerjali, a tributary that enters the Manu about halfway between the research station and Tayakome. The expressed motive for the move was to be closer to the research station and the source of goods it represented.

Thus began a long and satisfying interaction between the scientists at the Cocha Cashu Biological Station and the three families of Cumerjali, particularly those of Zacarías and Sandoval. Showing entrepreneurial spirit, they planted gardens larger than necessary for their own needs. For nearly twenty years, they have supplied the station with manioc, bananas, plantains, papayas, avocados, hot peppers, and occasionally other produce. In return, we have traded axes, machetes, mosquito netting, cloth, needles, pots, matches, flashlights, batteries, fishhooks, nylon monofilament, and sundry other articles on request. We eat well, and their wants for manufactured goods are mostly satisfied. It is a good deal all around.

As our trade relationship with the Machiguengas has flourished, so has it brought change to the lives of our counterparts. A new generation has grown up wearing manufactured clothing, listening to music on the radio, and speaking Spanish. As these and other attributes of Western culture are acquired, the traditional oral history and knowledge of medicinal plants are being lost. The new generation has no desire to live in the old way, yet its members are not well prepared to compete in the outside world. They are caught partway in a transition to modernity and are apprehensive about what the future will bring. Their well-being rests in unimpeded access to the land and its resources.

The Machiguengas and other indigenous groups that occupy the park's heartland regard their presence in the park as a birthright, not a special privilege granted by people of another culture who come from far away. By long-standing practice, indigenous people have been allowed to live within the park and to support themselves by farming, fishing, and hunting. Everyone else is prohibited from engaging in any of these activities within the park.

The presence of people within the park thus creates a set of double standards, and in part it is these double standards that threaten the park's future. Residents are allowed to clear land, plant gardens, and construct settlements, but they are not allowed to use firearms or to engage in commercial activities based on the park's resources. A prohibition against motor-driven boats and other mechanized devices has been honored in the breach because the teachers who live at Tayakome and Yomihuato both have boats and motors supplied by the Dominican mission in Shintuya. The system of double standards has served the park's purposes over the short run, but it is certain to fail over the long run, for it serves only to impose second-class status on the park's inhabitants.

One by one, the Machiguengas of Tayakome, Yomihuato, and Cumerjali are seizing opportunities to journey outside the park, to Boca Manu, Shintuya, or even Cuzco. They return with a transformed view of the world and of themselves. They discover that there are other Machiguengas who live outside the park and who are free to use guns and exploit the natural resources around them for profit. Through such contacts, they are beginning to see themselves as an underclass, and resentment is building.

This awareness and the resentment that comes with it are being politicized by activists belonging to Peruvian indigenous rights organizations. The people of Tayakome and Yomihuato are being encouraged to stand up to the park's administration and claim their "rights." Among the rights being sought are legal title to community lands within the park and freedom to engage in commerce with the outside world. Ambitious leaders within the two communities are taking up the drumbeat as they seek to promote their own status and fortunes.

It will not be long, I predict, before the people of Tayakome and Yomihuato attempt to assume control over their own lives. When that happens, the administration of the park will be under intense pressure to make concessions at the expense of nature.

Under the concept of a national park as codified in U.S. law, parkland is viewed as a place reserved for nature, a place where humans are permitted as visitors but not as permanent residents. The concept has worked well in the United States, but it is problematic in many tropical countries, where indigenous people often occupy the entire landscape. Herein lies a moral dilemma. Should tropical countries forgo parks because there is no convenient place to locate them? Or should parks be redefined as places where nature is intended to co-exist with people? In practice, many tropical countries have adopted the latter definition.

Many conservationists have been willing to go along with the co-habitation concept, asserting that indigenous peoples have been part of the scene for thousands of years without fundamentally altering the ecology. Although that assertion itself is arguable, the tacit assumption that the past will extrapolate into the future is far more arguable. Technology has fundamentally altered the footprint human beings leave on nature. A culture that hunted with bows and arrows in the past will certainly hunt with firearms if given the opportunity. Study after study has shown that once a premodern society acquires firearms, it overexploits the game supply.[1] The same argument could apply to the substitution of chain saws for stone axes. Moreover, in the case of the Machiguengas, minimal availability of modern medicine has unleashed a demographic explosion. Firearms, chain saws, and a demographic explosion are not the stuff of peaceful coexistence with nature. Hence my skepticism for the argument that parks and people can coexist in the future because coexistence in the past has not yet brought about destruction of the environment.

Earlier, I mentioned the breakaway group of Aguilar, Zacarías, and

Sandoval, the three Machiguenga heads of family who moved closer to the research station to facilitate a trade relationship with the scientists. As soon as we were aware that the station had neighbors, we recognized the threat their presence posed.

At its founding, the research station was located in the center of an unpopulated region. Scientists at the station consequently profited from a rare opportunity to study one of the few spots in the entire Amazon basin of Perú that was not affected by hunting. Where there is no hunting, animals may lose their fear of humans and become "habituated." Habituated animals can be observed at close range, and they have been the subject of much of the research conducted at Cocha Cashu. Habituated subjects allow the gathering of detailed information on diet, ranging pattern, social behavior, and so on. Among the animals to have been habituated at Cocha Cashu are eight species of primates, giant otters, razor-billed curassows (a turkey-sized game bird), white-winged trumpeters, and others. The pivotal role played by habituated animals in our research made the maintenance of a strict no-hunting zone around the station a make-it-or-break-it issue for the scientists.

Shortly after the three Machiguenga families moved downstream to be nearer the station, we received a visit from Zacarías. At the time, a student from Princeton University, Charles Janson, was conducting research on capuchin monkeys for his doctoral dissertation. The day after Zacarías's visit, Charlie was distraught to discover that three individuals were missing from a group of monkeys he had been observing closely. That same day, he discovered a Machiguenga arrow near the river, in an area the capuchins had been using regularly.

The next time Charlie met Zacarías, he confronted him angrily, asserting that Zacarías had shot three capuchins. Charlie then broke the arrow over his knee to demonstrate the intensity of his feelings. Zacarías did not deny that the monkeys had been shot, but he protested that it was not he but his son-in-law, Cornelius, who had done the deed. Even so, Zacarías was obviously unnerved by Charlie's knowledge that precisely three monkeys had been killed—evidence, perhaps, that we possessed supernatural powers. Moreover, Charlie had been enraged, and he had made it plain to Zacarías that if he and the other Machiguengas in his group wanted to have a trade relationship with the scientists, they would have to respect a no-hunting zone around the station and extending several kilometers upstream.

Concern arising from the incident prompted us to ask the park administration to support the no-hunting zone. Predictably shying away from any whiff of controversy, the director replied that it was not in his power to implement a no-hunting zone because, as lawful residents of the park, the Machiguengas were entitled to hunt anywhere they wished. By default, it fell to us to negotiate with our neighbors.

Eighteen years have passed since Cornelius killed the three capuchins, and to my knowledge there have been no breaches of the agreed-on no-hunting zone. Zacarías, Aguilar, and Sandoval are all welcomed as friends by residents of the station, and there is no tension between us of which I am aware. Nevertheless, I doubt that the status quo will endure much longer.

Those eighteen years have seen the gradual repopulating of the main trunk of the Río Manu. Where once there were three families living near the mouth of the Río Cumerjali, there are now ten, seven of which have settled along the Manu downstream of the Río Cumerjali. The seven new families are headed by the sons and sons-in-law of the original three founders. Aguilar is now an old man, but Zacarías and Sandoval, long since grandfathers, both still produce offspring at the rate of nearly one per year. The settlements swarm with children, most of whom survive to adulthood because they have access to public health services provided by the government. Where there are now ten families, in another eighteen years there will be twenty—that is, if nothing else changes, which is unlikely.

Over the past eighteen years, change has come not only to the park but also to the surrounding region. As I write, helicopters pass over-head, ferrying fuel to an exploration crew for the Mobil Corporation, their rotors creating a slapping sound not heard before in this far corner of the Amazon. Mobil's concession to the north of the park is in one of the wildest places remaining on earth. There are no roads or settlements, not even a mission or an airstrip. It is true wilderness, occupied only by uncontacted people of uncertain ethnic affinities.

The rude and uninvited intrusion of Mobil into a region that has known nothing but absolute isolation since the days of Fitzcarrald is bound to instill panic. Terrified people are likely to respond in one of only two ways. Either they will flee or they will attack. If they flee, there is a good chance that some of them will go south and into the park, where they will encounter other indigenous groups. Encounters between alien native groups are instinctively hostile, as all interactions between Yaminahuas and Machiguengas had been before the Yamina-huas made contact.

Mobil officials reassuringly assert that they are taking every reasonable precaution to avoid incidental encounters with indigenous people, but such assertions are little better than wishful thinking, for it is the Indians, not Mobil, who will decide whether there are to be encounters. And if the Indians make contact, it will not be friendly. What then? How would Mobil's executives respond if a company work crew were to come under attack? Would the quest for petroleum yield to the threat of further attacks? Not if history is any indicator. In the past, under such circumstances the government has been called on to provide troops to protect corporate workers.

Shell International Petroleum Company also holds a concession in the region. The company has discovered a major deposit of natural gas near the western boundary of Manu National Park. Reportedly, the deposit contains enough gas to satisfy Perú's energy needs for the next seventy years. Shell is currently drilling wells, and a pipeline for transporting the gas to the coast is being planned.

The native groups that happen to occupy the Shell concession are better off than those in the Mobil concession. Many have been contacted and can communicate to some degree in Spanish. Shell has hired onto its work crews a number of Machiguengas and some of the same Yaminahuas who once lived in the park. Most significantly, Shell has been obliged to negotiate the terms of its production operation with the groups that hold legal rights to the area (known as the Nahua-Kugapakori Reserved Zone), most of whom are Machiguengas. The most fervently expressed desire of the Machiguengas was that Shell prevent their lands from being overrun by *colonos* from other parts of the country. Shell has reportedly agreed to restrict the entry of unauthorized people into its concession.

Jarring change awaits the inhabitants of the Shell concession regardless of the terms they negotiate. Drilling operations and pipeline construction will bring cargo aircraft and heavy equipment into a hitherto quiet corner of the Amazon's headwaters. With the skilled workers imported to develop the field will come demands for women and liquor. The pacific and relatively passive Machiguengas are not prepared for changes as rude as these. Their reactions cannot be predicted, but should any of them be repelled by the oil culture, the one place they will view as a peaceful refuge is Manu National Park.

No one knows how many indigenous people now inhabit the park. I am aware of at least 500 contacted Machiguengas and one village of contacted Huachipairis. In addition, there are several hundred uncontacted Machiguengas and unknown numbers of uncontacted Mashcopiros and Yaminahuas. The total is probably less than a thousand. Still, that is a lot of people to be living in a national park.

Given the rapidity with which the population of contacted Machiguengas is growing, the virtual certainty that first contact will be made with additional groups, and the possibility of an influx of refugees from the Shell and Mobil concessions, the park faces a demographic explosion for which it is completely unprepared. At present, the park's administrators are opposed to demarcating indigenous zones within the park, although this idea is being pressed by some indigenous rights organizations. Zoning the park into indigenous and nonindigenous areas would leave it looking like a piece of Swiss cheese. And continuing population growth would generate enormous pressure to expand the indigenous areas at the expense of the rest of the park.

What is the alternative if zoning is ruled out? In the absence of any stated policy, by default the policy becomes one of looking the other way while events run their course. Sadly, I must conclude that this is what appears to be happening.

Is there a solution? There is. What I propose is a carefully constructed and voluntary relocation program built on the manifest desire of contacted indigenous groups to acquire goods and an education for their children and to participate in the money economy. Bringing about such a program will require a political courage I have not seen among the responsible authorities. Yet I believe that a well-designed program of incentives could entice some residents to leave the park voluntarily. Unclaimed land is available outside the park's boundaries along the left bank of the Río Madre de Dios. The available land is uninhabited, is of a quality comparable to that currently occupied by the park's residents, offers unexploited game populations, and lies along a major river where boats pass daily.

Many families might be drawn to the promise of title to the land, a school, and a public health post. Members of the younger generation could be lured by opportunities to work for pay, especially in the rapidly growing ecotourism industry. They would not go in a rush, however. An adventuresome vanguard would have to pave the way psychologically for others to follow. Members of the older generation, unable to speak Spanish and accustomed to the ways of the past, would overwhelmingly choose to remain within the park. But because they are largely beyond their reproductive years, members of the older generation would not constitute a demographic threat to the park.

Clearly, a relocation program would not solve the park's people problem overnight. If it were conducted entirely through voluntary choice, a generation might pass before the park's human population stabilized. As the more worldly and ambitious young people moved their families to the outside, recently contacted groups from the head-waters would seek to ease the transition to modern life by living within the park along the main course of the Manu River.

Two enormous obstacles stand in the way of the implementation of such a relocation program. Both are political, but one is short-term and the other long-term. In the present political environment of Perú, indigenous rights are being championed by many groups, including the government. Education and public health services are being car-ried to remote villages where such things were previously unknown. Numerous private organizations and religious groups are pressuring the government to increase the area of land titled to indigenous groups. In this political climate, proposing the relocation of an in-digenous population would be analogous to advocating racial segre-gation in the United States. The topic itself is so politically incorrect that it cannot be rationally discussed.

Over the longer term, the successful conduct of a voluntary relo-cation program, could one be implemented, would require a continu-ity of both policy and financing that would be unprecedented in Perú. Given the high rate of population growth in the country as a whole and the accelerated pace of economic change under the Fujimori gov-ernment, Perú will be a radically different place in twenty-five years. The thought of patiently pursuing a politically sensitive policy through a prolonged period of rapid social and economic change sim-ply stretches the imagination.

In the relatively brief period during which I have known the park, one village, Tayakome, has spawned three communities. Soon these communities will fission and additional settlements will appear along the river. I fear that in only a few more years, the no-hunting zone around the research station will become impossible to maintain. That will be the signal that my years in the Manu have reached their end.

But there will be no such decisive end to the park. Gradually, it will lose its wilderness character. Gradually, the animals will become scarcer and wilder. A mythical reputation will continue to draw tourists for a time, but eventually word will get around that the Manu is not what it used to be. What then?

Having lost the luster that made it one of the world's premier rainforest parks, and swelling with an ever larger and more assertive indigenous population, the Manu will imperceptibly pass from being a national park to being a reserve for its indigenous inhabitants. But no one will ever know when it ceased being the one and started being the other. A cynic might say that it has happened already—that the park never really had a chance.

5

PARKS: THE LAST BASTIONS OF NATURE

WITH FEW EXCEPTIONS, Manu National Park being one, existing parks in the tropical forest realm are a sorry lot. Few are large enough to maintain healthy populations of top predators. A majority exist only on paper, having no staff whatsoever. Most are poorly designed, the boundaries having been drawn in such a way as to make them indefensible against encroachment. A great many have people living within them. A significant number are already seriously degraded by illegal activity. Some no longer exist in a biological sense.

Despite its gathering problems, the Manu is better off than most parks in the developing world. By the beginning of the twenty-first century, many parks will be lonely islands in a human-dominated landscape. Some already are. Take Nairobi National Park in Kenya, for example. When a visitor looks out over the savanna, what greets the eye is a picture-book slice of wild Africa replete with wildebeest, zebras, gazelles, and buffalo. But if the visitor turns around, the view is a chain-link fence behind which is the skyline of Nairobi, with its high-rise office buildings and apartment towers. Nairobi National Park is ahead of its time, a park of the future. It is hardly wild Africa; it is a glorified game ranch. Its existence on the fringe of an expanding urban area is a glaring anachronism maintained for the tourist dollars it brings to the government.

As of 1996, approximately 3.7 percent of the earth's land area had been formally designated as parkland (IUCN Categories I–III) by the IUCN (World Conservation Union, formerly the International Union for the Conservation of Nature and Natural Resources). The organization has declared a goal of 10 percent, but that is a highly optimistic figure. Realistically, the eventual total will probably be less, doubtfully more than 7 percent to 8 percent. Within this global picture, tropical forests have been accorded slightly above-average treatment. As of 1997, 8 percent of the world's tropical forests enjoyed protected status, although much of this was in paper parks.[1]

Given the disproportionate biological richness of tropical forests, the level of protection they have been afforded is disappointing, amounting to the proverbial bone thrown to a dog. But the prospects for improvement are not bright. Creating new parks means incurring substantial opportunity costs at a time when the Tropics are under demographic siege. National government officials know this, even if they shrink from admitting it publicly. They are consequently reluctant to commit land to protection when they know it will be needed by hungry citizens in the not-too-distant future. Moreover, primary tropical forests constitute rich troves of unexploited resources. Business interests in every country are pressuring governments to make timber resources available for exploitation. Mostly, it is foreigners who are clamoring for more parks, not the citizens of the countries themselves. In the minds of government officials, there are already enough parks to serve the needs of foreigners. Why should they want any more? Most do not, as the statistics show.

The rate of creation of protected areas in the world's tropical forest regions peaked in the late 1980s. Since then, the rate of establishment of new areas has dropped precipitously. Unless new park foundings unexpectedly surge in the near future, less than 10 percent of the tropical forest biome will ever be protected. Now, consider that only about 7 percent of the world's land surface is covered by tropical forest. Putting the two figures together, we see that conservation science is faced with the Sisyphean task of perpetuating more than 50 percent of the world's biodiversity in 0.7 percent of the earth's land area. If saving biodiversity were not such a serious matter, this prospect could be regarded as a bad joke.

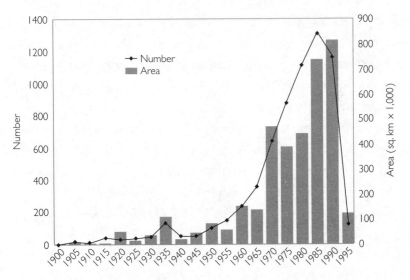

Creation of protected areas in tropical forest regions.
Copyright © World Conservation Monitoring Centre.

How large should a park be to sustain its biological richness for the foreseeable future? The short, but impractical, answer is "As large as possible." As a rough guide, the area should be the size needed to sustain genetically viable populations of top predators.

In the Amazon basin, the top predators are pumas, jaguars, harpy eagles, and giant otters. Nothing is known about the ecology of pumas in the Amazon, so this predator must be omitted from my calculations. But the needs of the remaining predators are quite telling. A female jaguar requires a minimum of 20 square kilometers for herself and her cubs.[2] The territory of a pair of harpy eagles encompasses some 50 square kilometers.[3] And a family of giant otters needs one to several oxbow lakes along a major white-water river. The 2 million–hectare Manu Biosphere Reserve (the core park plus adjacent buffer zones), for example, supports 10 families of giant otters, a total of about 60 individuals.[4]

To determine the minimum area needed to support a genetically viable population of top predators, we must find the area required to support 300 breeding females of each species. Why 300? Because geneticists calculate that a population will lose genetic diversity unless it contains at least that many.[5] In the case of the jaguar, one needs to multiply territory size (approximately 2,500 hectares) by 300, which gives a minimum required area of 750,000 hectares. For the harpy eagle, the minimum area is 1.5 million hectares. And for the giant otter, it is an unattainable 60 million hectares. Unless the giant otter can tolerate inbreeding, it simply cannot be effectively protected in any existing reserve. Taking the known spatial requirements of jaguars and harpy eagles and splitting the difference between them produces the round number of 1 million hectares.

The numbers bring my argument into sharp focus. Of the 117 nature preserves situated partly or entirely within the Amazon basin, only 18 exceed 1 million hectares, another 60 contain between 100,000 and 1 million hectares, and the remaining 39 are small.[6] Compared with other tropical areas, Amazonia represents the best-case region because it remains largely roadless and undeveloped.

The Indo-Pacific region, an area much larger than tropical South America that includes all of southern and southeastern Asia plus the tropical Pacific Ocean, contains a larger total number of reserves (1,185), but on average they are smaller.[7] More than half include less than 10,000 hectares, and only 122, about a tenth of the total, exceed 100,000 hectares. Yet Asia harbors far more large, space-demanding mammals than does South America, including tigers, leopards, elephants, rhinoceroses, tapirs, sun bears, and others. So on the basis of reserve size alone, the prospects for biodiversity conservation in Asia are far from ideal.

Inadequate staffing also affects many tropical parks. As an example, consider the contrast between nature preserves in the Brazilian Amazon and parks in the United States. On paper, Brazil maintains 30 nature preserves in the Amazon, but only 10 of these employ even one guard (1995 data). All 30 preserves together employ a total of 23 guards; in contrast, in the United States the National Park Service employs about 4,000.[8] Each guard in the Brazilian Amazon must police more than 6,000 square kilometers, an area larger than the state of Delaware. Such understaffing reflects the low priority placed on nature protection by the government, notwithstanding the fact that Brazil is one of the wealthier tropical forest countries. By comparison, the density of guards in parks in the United States is more than 70 times higher than that in Brazil.

Underscoring the impossibility of the task they are supposed to perform is the fact that Amazonian park guards are unarmed and lack authority to arrest violators. All they can do is submit reports to superiors, who are often far away in regional capitals. Enforcement suffers from disinterest and lack of cooperation on the part of police agencies and a procedural vacuum resulting from a lack of legal precedent. Park guards routinely experience such debilitating working conditions throughout the tropical world; in many places, enforcement of park regulations is virtually nonexistent.

Not surprisingly, the upshot of such institutional weaknesses is that preserves in Amazonia and elsewhere are being violated on a large scale. All 30 Brazilian preserves have poaching problems, 23 have been invaded by squatters, logging activity affects 18, land titles are being disputed in 18, 13 are being degraded by mining, and 8 harbor indigenous populations.[9] And this is in the Amazon, where the pressure on resources is substantially less than in other tropical regions.

Whereas logging, mining, and some other illegal activities conspicuously alter habitat and can be detected by aerial surveillance, some parks are under the cloud of an invisible threat no guard can deflect: encumbered title to the land. Only 12 of the 30 nature preserves in the Brazilian Amazon lie wholly within the public domain, and most of these 12 are small ecological stations. Fifteen of the region's 16 national parks and biological reserves (the largest conservation units) are compromised by private inholdings. The cost of buying out the owners is considered prohibitive by the government.

Private ownership of Amazonian lands also limits the potential for future growth of the region's nature preserve system. Private holdings and claims covered 65 percent of the Brazilian Amazon in 1990, notwithstanding the fact that much of the area was still trackless forest occupied only by indigenous people, whose own claims to the land are largely unrecognized. Land in the Brazilian states experiencing the most rapid development is already at least 83 percent in private hands (for example, in Maranhão and Acre) if not already entirely so (Mato Grosso, Tocantins).[10] The Amazon is no longer the limitless frontier it was just a few years ago, such is the pressure to stake claims on unexploited resources.

Another problem is that many parks offer ready access to encroachers. Park boundaries are often drawn for convenience or to accommodate competing claims on the land, with little thought given to defense, despite the anticipation of meager to negligible staffing. Such easy access only encourages poaching, mining, removal of logs, and other infractions. If poachers will walk ten kilometers beyond access roads in search of game, as is likely, then 75 percent of the areas of existing nature preserves in the Brazilian Amazon are accessible to intruders.[11]

At this point, I would like to recount a personal experience with a paper park, for it illustrates many of the points made in the previous paragraphs.

Early in 1985, I traveled to the northwestern part of Perú to visit Cerros de Amotape National Park. I carried with me a formal authorization issued by the Ministry of Agriculture in Lima, the division of the Peruvian government that administers national parks. Without authorization, I could not legally enter the park. My companions and I arrived in the regional capital, Tumbes, late one afternoon. The next morning, after considerable inquiry, we tracked down the local office of the Ministry of Agriculture. There, we introduced ourselves as scientists and presented our authorization document, explaining that we wanted to conduct a study of birds in Cerros de Amotape National Park.

The administrator was baffled. The purpose of the authorization was a mystery to him. He called in two subordinates. Had either of them heard of Cerros de Amotape National Park? Yes, one of them said, he knew about it. He had never been there, but knew that it began about twenty kilometers out of town. He then went to a cabinet and pulled out a torn and yellowed map of Tumbes. On it was a line indicating the boundary of the national park. As we studied the map, the one official who knew of the park informed us that to his knowledge, we were to be the park's first visitors. At the time, the park had formally been in existence for more than a decade.

How should we proceed to reach the park, we inquired, and once there, how would we get into it? No one in the office had answers to these questions because none of them had ever been to Cerros de Amotape. At this point, the official who had produced the map offered another suggestion. Someone he knew at the ministry had been to the park. His name was Eulogio. He wasn't at the office that day, but perhaps we could find him at home.

We finally located Eulogio, and he turned out to be a good-natured fellow who seemed to enjoy his newfound status as an expert on Cerros de Amotape. We were surprised and gratified when his boss assigned him to us for ten days so he could guide us to the park. We hired a pickup truck, loaded our things into it, and drove out of town on a potholed dirt track. The road ended at a farmhouse, still several kilometers from the park boundary. Eulogio knew the owner and knew he had some pack animals. After some bargaining, we settled on a price and loaded several mules and donkeys with our food and camping equipment.

An hour or two later, after trudging along in the hot sun, we reached a line of woods that Eulogio allowed was the park boundary. The trail continued on with no indication that we had entered hallowed ground. The owner of the pack animals was familiar with the route, but to him, the land was just backcountry, open to any and all. Only educated foreigners like us would know we needed a written authorization from Lima to enter.

Such formalities were not a concern to local residents. Herds of goats ravaged the vegetation near the boundary. The local army detachment, we were told, helped itself to the forest for building materials and fuelwood. Cattle ranged freely into the farthest reaches of the park. Spent cartridges told of hunting past.

We passed about a week in Cerros de Amotape, camped beside a clear-water stream that cascaded through a picturesque canyon over water-sculpted rock formations. The spot was a lovely slice of green in an otherwise drab and leafless landscape. Yet biologically, Cerros de Amotape is anything but drab. It lies in the middle of a biological hot spot, a center of endemism that is home to more than twenty species of birds that occur only in a small area of southwestern Ecuador and northwestern Perú. I am told that the Ecuadorian side has been completely deforested.

Cerros de Amotape National Park and two adjacent areas of lesser conservation status constitute Perú's Northwest Biosphere Reserve. The reserve is the last bastion of birds, plants, and other organisms endemic to the Pacific dry forest of South America. It is a world-class haven of biodiversity. But if the reserve attracts only one group of visitors every ten years, how long will it continue to exist? If the people who are nominally responsible for administering it have never been there, how can it be protected?

In 1985, the situation of the Cerros de Amotape was not yet grave because the park contained no people as permanent residents. As mentioned earlier, in some countries (e.g., India), the presence of humans within park boundaries is an accepted and acknowledged fact. Residents are expected to respect the wildlife but are otherwise permitted to go about their lives in a normal fashion, planting crops and grazing domestic stock.

In many parts of the Tropics, it is simply impossible to delineate a million-hectare tract of real estate that does not include the traditional lands and settlements of indigenous people. Even when the enabling legislation describes an area as a strict nature preserve and explicitly prohibits hunting and other use of natural resources, the presence of indigenous people is tacitly accepted as a fact of life. It is easy to rationalize natives as being "part of the ecology" of a park and not a threat to its integrity. But as we have seen in the case of Manu National Park in Perú, such rationalizations are a form of self-delusion.

Unauthorized human residents of parks are often referred to as squatters. Squatters are easily distinguished from indigenous inhabitants because typically they were born outside the protected area's boundaries. In Latin America, it is not uncommon for land to be usurped prior to the formal establishment of a park, as soon as it is public knowledge that the area is destined to come under protection. Claims of prior residence are then often tacitly honored by park authorities. Yet turning a blind eye to squatters only aggravates the threat because the presence of people almost inevitably attracts more. Many parks are quietly invaded by squatters who furtively trickle in around remote or unguarded sectors of the boundary. Invasion by squatters obviously threatens the integrity of any park; yet, for reasons to be discussed, squatters are rarely expelled.

The notion that parks should provide a home and a livelihood for human residents conflicts with the concept of a national park that is familiar to most residents of the United States. Under the statutes applied by the National Park Service, a national park is a place reserved for nature. People are not allowed to reside within parks, with exceptions made for authorized park personnel and concessionaires. Members of the public are permitted to enter only on a temporary basis, as benign visitors. All commercial activity is proscribed, apart from the operation of licensed concessions. Unauthorized removal of any biological or mineral product is prohibited, although collection of scientific specimens can be authorized if adequately justified. Even the picking of wildflowers and the feeding of wildlife are prohibited. By and large, citizens are accustomed to these rules and obey them.

In other parts of the world, parks exist under a wide range of definitions. For a protected area to qualify for listing under the highest

IUCN categories, its regulations must conform closely to those just cited. In practice, however, status is often granted with one eye closed to irregularities.

Tropical parks tend to be established in remote frontier regions where there is unencumbered title to the land (disregarding claims of indigenous peoples). The very fact of remoteness justifies scant or no commitment to protection, a park on the cheap. Remoteness also ensures that tourism is minimal, a fact that deprives many parks of any constituency.

Forgotten, unguarded paper parks can experience progressive degradation that begins imperceptibly and then picks up momentum as population builds up around the boundaries. First, a road into the region is built. Prime timber resources are quickly exhausted along the road, at which point the park becomes a magnet for loggers. Mineral deposits are discovered, and miners move in. But mounting pressure on the park's resources brings no response because there is no government presence. The first incursions amount to nibbling around the borders. When it is perceived that illegal activities can be carried out with impunity, the nibbling can turn into a full-scale assault. Logging, hunting, mining, and land clearing can coalesce into a multifronted onslaught. At the end of the process, there is nothing left of the park but its name. Sadly, the integrity of parks is undergoing progressive erosion in many parts of the world. Nearly every park in the Tropics is being degraded, some rapidly, some slowly, but few are immune to the relentlessly growing pressure of human demands on the environment.

Parks are the last refuge for much of the world's biodiversity, yet throughout the Tropics they are crumbling under the assault on their natural resources. To illustrate the many ways in which tropical parks are being degraded, I present here a selection of vignettes from around the world. Our odyssey will begin in the New World; then we will hop to Africa and conclude in Asia. In general, none of this information has been systematically organized or published. There are a lot of reasons for that. People do not want to hear more bad news about conservation. National governments are sensitive to information that can be interpreted as critical. Conservation organizations shrink from publicizing their failures. Many of the observations that follow, however objective, can be construed as politically incorrect. But no one ever claimed that conservation in the developing world would be easy.

NEW WORLD TROPICS

Nevado de Colima National Park, Mexico
Volcán Colima is a spectacular volcanic mountain set in majestic conifer forests and capped with a glistening, snow-clad pinnacle. In the winter of 1983, I took a Princeton University field ecology class to this biologically rich corner of Mexico. At the bottom of the mountain, a sign beside the road informed us that we were entering the national park. On the way up, we passed several loaded logging trucks, air brakes hissing as they gingerly eased their way down the steep grade. We camped in an open pine forest near the top of the mountain, where we had a breathtaking view of its summit.

The next morning, as we began to gather observations on migratory birds, I noticed that we were in a strange place. It was an alder forest, but I had never seen such alders before. I was accustomed to the shrubby ones that grow along streams farther north, but these were immense, some of them a meter and a half in diameter. What caught my attention was that the trees were widely scattered, with brambles and other brush crowding the gaps between them. Winding through the brush were cattle trails, profusely littered with cowpies. I soon realized why there were such wide gaps between the alders. The cattle were eating everything but the brambles and brush, preventing any regrowth of alders or other trees. We had arrived in time to bear witness to a forest that had joined the living dead.

Guatemala

A comparison of aerial photographs taken in 1982 and satellite images from 1991 revealed that nearly half the forest in Guatemala's Sierra de las Minas Biosphere Reserve had been destroyed in the intervening period. The deforestation resulted from expansion into the park of agriculture and logging from already exploited adjoining lands.[12]

Cerro Campana National Park, Panama

Only a short drive to the west of Panama City on the continental divide is a park that could provide welcome weekend relief from the city's stifling heat. Winding up from the coastal highway, the dirt access road passes through cattle pastures and coffee plantations. Finally, near the continental divide, the road passes a large sign welcoming visitors to Cerro Campana National Park. But there is no entrance gate. The road simply continues through fields and more coffee plantations. When I visited the park in 1970, it was clear that the land had been invaded and occupied. No one was there to hold the line.

Tayrona National Park, Colombia

The following excerpt from a 1995 article in the *New York Times* speaks for itself:

Shortly after 7 o'clock on a September evening, Hector Vargas, the head warden of Tayrona National Park, was driving along a main road with two colleagues when several shots were fired. Hours later, Mr. Vargas was dead and one of his assistants had suffered permanent brain damage.

He was the sixth park official in Tayrona to be killed in the last 20 years. Mr. Vargas's colleagues can only speculate about which of the myriad threats to the park's survival—illegal settlers, entrenched landowners, drug traffickers, wood cutters, Indian treasure hunters or armed mercenaries—might have killed him. But his death exposes the perilous circumstances under which the country's 43 national parks are fighting for survival. One week after Mr. Vargas was shot, an official at Sierra de la Macarena Park in eastern Colombia was murdered. Days before, the authorities temporarily closed Los Nevados, a national park in southwestern Colombia, because of a forest fire that they attributed to arson.

Government officials and environmental activists say that the country's national parks are protected areas on paper only. An estimated 20 percent of the country's 22.2 million acres of park land is in the hands of illegal settlers or private owners. The Government prohibition against private ownership in national parks is often ignored by local officials who give deeds to people. Reserves such as La Paya, which covers more than one million acres along Colombia's borders with Perú and Ecuador, serve as refuges for rebels, as well as sites for growing coca crops and processing cocaine.

"Most of the parks are a reflection of the violence and social problems that face Colombia at large," said Cecelia López, Colombia's Environmental Minister. "The ecological pillage that we are witnessing is incredible."

Colombia is home to about 10 percent of the world's flora and fauna and is second only to Brazil in diversity of native and nonnative species. However, with sparse funds and an average of one park warden per 93,000 acres, the Government's Institute for Natural Resources, or Inderena, has been ill-equipped to protect the country's natural wealth. . . .

Over the last 30 years, landowners and illegal settlers have replaced trees with crops of plantain, sugar, maize and fruit. Hunters, wood choppers and gold diggers have left craters and tree stumps in their wake. A once-thriving Indian community has been reduced to one family living in an isolated area of the park. Many of their ancestors' graves have been looted. Park officials' attempts to stop such destruction often meet with violence.

When Tayrona was declared a national park in 1964, 90 percent of it belonged to the Government. Today, 80 percent of the park is in private hands, and Inderena lacks the resources to buy all but a fraction of the land. Deforestation and the more than 70,000 tourists who visit Tayrona every year have created water shortages and sanitation problems. . . .

"I was nothing compared to these powerful forces," said a former head of Tayrona National Park who asked not to be named and who received death threats during his tenure.[13]

Ecuador

Confusion reigns in Ecuador over the legal basis of protected areas, and as a result, different government sectors can authorize incompatible activities, for example, oil exploration. When oil was discovered in Yasuni National Park, the country's largest, the government promptly signed commercial agreements that allowed a major new production field inside the park. As recompense, the government promised that an equivalent area of parkland would be created somewhere nearby. During the same period, discovery of oil, along with the clearing of forest for oil palm plantations, prompted an official change in the borders of Cuyabeno Wildlife Reserve as well.[14]

Perú

The mere declaration of intent to create a new reserve can unleash a resource rush, demonstrating the perverse logic that can prevail in some developing countries. I saw this happen in Perú after the government announced the creation of the Tambopata-Candamo Reserved Zone. Under Peruvian law, a reserved zone is land in limbo. The status signals that the government is holding development in abeyance while alternative uses are debated. It was the hope of conservationists in Perú that declaration of the reserved zone would be the prelude to making much of the area a national park.

However, as soon as the declaration of the reserved zone became common knowledge in the region, scores of families of land-hungry *colonos* from the Andean highlands invaded the designated area to clear farm plots and establish themselves as residents. The reserved zone itself was not such a magnet to settlers. The attraction was the speculation, widely circulated among the *colonos,* that if the area became a park, the government would have to buy them out, thereby netting them a profit on land for which they had paid nothing.[15]

Bolivia

In a recent conversation with nature photographer André Bärtschi, I heard a similar story emanating from Bolivia. The Bolivian government had authorized a new park, the Madidi. Lying on the border with Perú, the area was designed to complement the hoped-for Tambopata-Candamo National Park, forming a huge binational protected area totaling more than 2 million hectares.

Bärtschi visited the formally declared, but not yet implemented, Madidi National Park in 1996. What he encountered was the equivalent of a looting. In the absence of any official presence, loggers had invaded the park in force and were systematically stripping it of mahogany. Market hunters had followed the loggers to supply them with game meat. Bärtschi was shocked at the sight of a day's bag of spider monkeys piled up in a mound.

These stories affirm the essential need for law enforcement in maintaining the integrity of parks. If vigilance is relaxed for even a moment, opportunists will take advantage of the lapse.

Here is a story to underscore the latter point, as told by a close friend, Joseph Wright, who has been on the scientific staff of the Smithsonian Tropical Research Institute (STRI) in Panama for nearly twenty years. One of his administrative duties has been to supervise the force of game guards that patrols Barro Colorado Island and the contiguous Barro Colorado Nature Monument, an area totaling 59 square kilometers. (Note that this represents one game guard for every 3 square kilometers, and recall that protected areas in the Brazilian Amazon have one guard for each 6,000 square kilometers.)

In December 1989, Barro Colorado Island was evacuated for five days and nights by order of the U.S. military commander following the U.S. invasion of Panama. When the island's staff was allowed to return, the entrails of two deer were found brazenly dumped in the middle of a trail.

Protection of the nature monument's wildlife amounts to thinly veiled warfare. Guards are well armed and equipped with infrared floodlights and night scopes for long-distance detection of small boats at night, when most poaching occurs. The force makes about twenty arrests per year, each involving one to several poachers. A poacher caught red-handed in June 1992 murdered agent Marcelino Castillo to avoid arrest. Fortunately, he was seen and recognized by another game guard and eventually imprisoned.

In a parallel effort to quell the poaching, STRI employs several people to conduct outreach programs in communities bordering the nature monument. Presentations at schools and other efforts to use persuasion to discourage poaching have so far failed to produce measurable results.

WEST AFRICAN TROPICS

The West African tropical forest is a zone of endemism, the exclusive home of such distinctive animals as the pygmy hippopotamus and the natty diana monkey. West Africa is also one of conservation's worst-case scenarios. In a recent conversation, J. Compton Tucker of the National Aeronautics and Space Administration's Goddard Space Flight Center commented that satellite images show only one sizable area of primary tropical forest remaining in all of West Africa. That one remaining tract is located in Guinea, in a region lacking any formal conservation status.

The national parks of West Africa are a shambles, as attested by eyewitnesses.

Senegal

One firsthand account came to me recently in a letter from a Peace Corps volunteer, from which I quote: "I would like to be involved in conservation in the tropics, even more so since being a forestry agent in Senegal, West Africa, while serving in the Peace Corps (1992–1994). I was able to see first hand the precarious imbalance between the human community versus the flora and fauna, or what was left of it. I was able to see 'Eaux et Forêts' signs still standing with a backdrop of blowing sand in what used to be a national forest, before the illicit charcoal industry got hold of much of it. I also saw the terrible representation of many government officials while certain precious trees were being hauled out from right under their noses."

Liberia

Former Duke University student Molly Docherty helped train and advise personnel of Sapo National Park in Liberia during her stint as a Peace Corps volunteer. After she finished her term and returned to the United States, civil war broke out and Sapo was overrun by rebel bands, who helped themselves to the park's timber and game. Logging was rampant during the war, with proceeds financing rival guerrilla groups. Even now that a nominal peace prevails, Liberian authorities have been unable or unwilling to halt illegal logging.[16] The scale of social upheaval caused by the war staggers the imagination. Most of the country was ransacked, 150,000 Liberians were killed, and 2.3 million people were displaced.[17] Sapo National Park, in effect, has ceased to exist.

Sierra Leone
Primatologist John Oates, a longtime friend and colleague of mine, has dedicated his professional career to studying the primates of West Africa. For more than a dozen years, he operated a field research center on Tiwai Island in Sierra Leone's Moa River. The central government in the faraway capital, Freetown, provided no support. To be confident that the habituated primates on which his and his students' research depended were not eaten by villagers, Oates had to make a deal with the local chief. Oates agreed to provide long-term employment to several villagers, in return for which the chief issued stern warnings that no one was to hunt the primates of Tiwai Island.

The informal arrangement worked for more than a decade, until the Liberian civil war broke out in 1990. Tiwai Island, only twenty-five kilometers from the Liberian border, was overrun by armed bands of warring parties. In the end, the habituated primates of Tiwai Island met their fate in the pot, despite Oates's best efforts to forestall the inevitable.

Côte d'Ivoire
The premier nature preserve in all of West Africa is Taï National Park in Côte d'Ivoire (Ivory Coast). Although the park was originally ga-

zetted (established) at nearly a million hectares, less than half the area within its boundaries remained forested in 1990.[18] Illegal logging has been one source of destruction, but the principal driving force has been the immigration of thousands of displaced farmers. Some of them are Ivorians who were forced out of traditional lands by gold miners, timber companies, and the creation of a hydroelectric impoundment.[19] However, many of the immigrants were extranationals from the Sahelian countries to the north, particularly Mali and Burkina Faso. Severe drought in the mid-1980s uprooted millions of Sahelians and induced them to migrate to better-watered lands in the south, making Côte d'Ivoire the involuntary recipient of 1.5 million refugees.[20] In equivalent terms, it is as if 30 million destitute aliens were to pour over the U.S. border in a nonmilitary invasion, swelling the ranks of United States residents by a number exceeding the combined populations of New York, Los Angeles, Chicago, and Philadelphia. Floods of Sahelians gravitated to the only unoccupied land they could find in Côte d'Ivoire: the forests of Taï National Park.

Ghana

The forests of neighboring Ghana are little better off. Primatologist Thomas Struhsaker traveled to western Ghana in 1995 to check on the status of three highly endangered endemic primates. There had been no reliable reports of any of the species since 1970, and it was not known whether they still existed. The best possibility for relocating them was in some reserves that satellite images showed were still forested.

When Struhsaker reached the reserves, he found that they had been conspicuously degraded by logging and fuelwood gathering. He also found them to be afflicted with empty forest syndrome, wherein habitat has been hunted to exhaustion and is devoid of animals larger than rats.[21] The continued existence of the reserves' three endemic primates is seriously in doubt.

By any reckoning, West Africa is a conservation disaster. What should be the response of the international community? A crash program to rescue some of the few degraded patches of habitat that remain, none of which is large enough to be biologically viable over the long run? Or abandonment of conservation efforts as a lost cause? The choice is not a facetious one, as conservation dollars are in short supply. Money spent in one region diminishes the amount that can be spent in another. The negligible cost-effectiveness of conservation dollars spent in West Africa to date suggests that scarce dollars are better spent elsewhere.

CENTRAL AFRICAN TROPICS

The evergreen forest zone of Central Africa lies to the east of the Dahomey Gap (a savanna that divides the forests of West Africa into two distinct sections) and includes parts or all of Cameroon, the Central African Republic, Congo, Gabon, Equatorial Guinea, and the Democratic Republic of Congo (formerly Zaire). Here is the African forest at its finest, supporting an extraordinary concentration of biodiversity, especially primates and other large mammals. The forest is home to gorillas, chimpanzees, more than thirty species of monkeys, forest elephants, and rare ungulates such as the bongo and the okapi. Human population densities in central Africa tend to be lower than in West Africa, so more of the primary forest remains. The inevitable assault of the timber industry on central Africa's forests has only just begun, hampered by political instability in several countries and an almost total lack of infrastructure.

I have visited two parks in the tropical forest zone of central Africa, one in Cameroon and one in Gabon. The situation of both was markedly better than reported here for West Africa, but only one of them appeared to be free from external pressures.

Cameroon

I visited Korup National Park in southwestern Cameroon with a small group of Princeton graduate students in 1989. Korup occupies a narrow sliver of land between the Ndian River on the east and the Nigerian border on the west. Our takeoff point was the town of Mundemba, a village lying at the edge of an interminable sea of oil palm plantations that now occupies much of the erstwhile tropical forest belt of Cameroon.

At the time of our visit, there were no organized facilities for tourists. We were allowed to camp for the night in an empty building that was part of a complex belonging to an oil palm plantation. The next morning, we met Ferdinand Namata, an experienced guide who had been recommended by a colleague. A representative of the World Wide Fund for Nature kindly shuttled us by Land Rover to the trailhead. We shouldered our packs, crossed the river on a rickety footbridge, and proceeded into the forest on a well-worn trail. Ferdinand was a superb guide. He knew the park and everything in it. I was happily surprised to discover that he knew the scientific names of many trees. Since most tropical trees lack recognized common names, we could rely on the language of science to communicate about the forest.

Korup is a beautiful park, preserving one of the finest tropical forests I have seen on any continent. The trees are tall and majestic, and the understory is uncrowded. One is never far from one of the fresh, clear streams that wind down the gentle grade toward the coast. Even better, there were few biting insects to detract from our enjoyment of the forest.

Every day, Ferdinand led us along different trails in quest of the park's wildlife, but the more time we spent in the forest, the more we realized how little wildlife there was. Occasionally, we could hear a group of panicked monkeys fleeing through the forest canopy, but in the week we were there, I failed to lay eyes on one. Ferdinand pointed out some old elephant sign but admitted that there had not been any elephants in that part of the park for at least two years. They had been poached for their ivory. The only animal I saw, and it was only the most fleeting glimpse, was a duiker (a forest antelope) that sprang across the trail one morning.

Korup, like the reserves in Ghana, is afflicted by empty forest syndrome. Its shape, a long, narrow slice of land, makes it impossible to protect. Hunters from Mundemba and other villages have free access to the park, having only to cross the river to gain entrance. Moreover, at the time of our visit, there were several villages within the park whose residents helped themselves to the unprotected game. But the most severe pressure on the park's wildlife came from outside Cameroon.

While walking the trails of Korup, we noticed little white mounds, usually atop trunks of fallen trees. Ferdinand explained that the white material was spent carbide, used to fuel the lamps of Nigerian market hunters who worked the park at night. We were many kilometers from the border, yet the white mounds were frequent. The long border with Nigeria was impossible to patrol. Moreover, how was a tiny staff of ill-equipped guards to arrest armed Nigerians at night? Better to patrol the forest by day, when it was safe, and to be home with the family at night.

Gabon

Relative to other African countries, Gabon has a tiny population. Compare its density of fewer than 5 persons per square kilometer with that of Rwanda, the highest on the continent, at 280 per square kilometer.[22] The hinterland is largely empty, thanks in part to government policies that encourage concentration of the population in towns, where services such as education and medical care can be supplied. The Gabonese people enjoy the highest standard of living in sub-Saharan Africa, attributable to an economy buoyed by oil revenues. Any Gabonese who so desires is guaranteed a government job. Much menial work is performed by imported laborers from neighboring countries.

So long as its oil reserves hold out, Gabon may offer the best hope for preserving biodiversity in tropical Africa. The country was almost entirely bypassed by the great wave of ivory poaching that spread across Africa in the 1970s and 1980s, so it still has a healthy population of elephants. The country also boasts 60,000 gorillas and 30,000 chimpanzees distributed over a large fraction of the national territory.[23] Lying in a major center of endemism, Gabon supports the richest primate communities on the planet.

I had the opportunity to visit Gabon in 1989 with a group of Princeton graduate students. The morning after our arrival, we took a taxi to the offices of the Ministère des Eaux et Forêts et de l'Environnement, the government agency that administers national parks in Gabon. As in many tropical countries where tourists are an oddity, written authorization to gain admission to a protected area must be obtained in the capital city. We were pleasantly surprised to find that the director was a genial gentleman whose office produced the required document that same morning, without a bribe. It was not the kind of service one expects in some African countries. Later, we heard through friends that the director had been jailed for his participation in an ivory smuggling ring.

Our destination was the Réserve de la Lopé, in the center of the country. La Lopé is a bright spot in an otherwise dismal picture of tropical forest conservation in Africa. The reserve does not have the status of a national park and has been partly logged for okume, a highly valued endemic tree. Otherwise largely intact at the time of our visit (1989), La Lopé harbored healthy populations of gorillas, chimpanzees, and elephants and was not under any pressure from poachers, miners, or loggers. How long such tranquility will endure remains to be seen. Gabon is a never-never land. Its economy is supported by oil and is subsidized by the French government. If the country had to live on renewable resources, it would be a different place.

La Lopé provided us with the thrill of seeing elephants, buffalo, and hundreds of forest primates that did not flee at the sight of a human. Colobus monkeys, mangabeys, and guenons of several species traveled unhurriedly through the canopy in large, noisy groups, in stark contrast with the situation at Korup. Apart from the wildlife, the best thing about La Lopé, in my opinion, was its director, Alphonse. Intelligent, educated, helpful beyond the call of duty, and above all dedicated to supporting the reserve, he is the finest model of a park administrator I have yet to meet in a developing country. If only other rain-forest parks had the benefit of such competence and integrity, the prospects for conserving tropical biodiversity would be more promising.

Madagascar

Often mentioned in the same breath with Africa is Madagascar, the world's fourth largest island. Located in the Indian Ocean roughly 500 kilometers east of Mozambique, Madagascar sits a world apart from Africa biologically. Evolutionarily isolated for 60 million years, Madagascar now harbors a unique flora and fauna, with nearly every organism apart from a few birds and bats endemic to the island. Madagascar is therefore on every conservation organization's list of priority areas for biodiversity. Restricted to the island are approximately 10,000 plant species, all the world's lemurs, dozens of chameleons and other reptiles, 200 frogs, and scores of freshwater fish, not to mention countless insects and other arthropods.[24]

The age and isolation of Madagascar have made it into an evolutionary museum, a repository of organisms whose ancestors disappeared from the rest of the world eons ago. Best known to the public are Madagascar's lemurs, small-brained but often colorful primates of diverse sizes and habits. The closest relatives of the lemurs, the Adapids, vanished 37 million years ago, but their next of kin, the lemurs, lived on in Madagascar in an evolutionary time warp, sheltered from modern primates, against which they could not possibly compete.

Madagascar was settled by humans about 1,200 years ago.[25] Quite remarkably, the nation's nineteen ethnic groups all speak a common tongue, Malagasy, related to languages spoken on the southern coast of Borneo, from whence the first settlers presumably came. Subsequent contacts with Africans across the Mozambique Channel, Portuguese traders, Protestant and Catholic missionaries, and, finally, a French colonial administration have left their marks on the culture.

The most biologically destructive of these cultural exchanges was the introduction of cattle from Africa. In some parts of the island, cattle are the measure of an individual's wealth. The point is not to eat them but simply to possess them—the more the better. Cattle are money on the hoof. Unfortunately, cattle are extremely destructive to vegetation. Not only do they devour the seedlings of trees and other native plants, but they also demand large quantities of forage.

Cattle forage is not abundant in a tropical forest. To expand the area that could be used by their cattle, early Malagasies set fires in the dry season to bring in fresh grass. The practice was so widespread and so devastating to native vegetation that by the time Europeans began to explore the country in the late nineteenth century, forest cover had disappeared over about 80 percent of the island's surface. Much of the interior plateau had become a vast wasteland. I have never seen a more desolate landscape—more thoroughgoing environmental destruc-tion—than on the central plateau of Madagascar. Large areas have eroded down to bedrock, which glistens in the sunlight as one passes over in an airplane.

When I first visited Madagascar, in 1984, it was the poorest country I had ever seen. In the near absence of a money economy, the govern-ment subsisted on foreign aid. Shops were bereft of goods. The filthy streets of downtown Antananarivo, the capital city, swarmed with urchins and beggars. Conditions were no better in the countryside. The people had no manufactured implements. Instead, they used rudimentary hand-forged tools. One was an ersatz machete, a narrow blade about 30 centimeters (1 foot) long, the end of which had been bent to form a loop. A stick jammed into the loop served as a handle. Another implement of similar construction but with a broader blade was used for planting rice, the staple crop. A lack of foreign exchange precluded acquisition of even the simplest manufactured goods. Not even in the poorest parts of the Dominican Republic, a low point in Latin America, had I found people too poor to afford a machete.

One part of the legacy of French colonialism in Madagascar was an excellent system of parks and nature reserves. Located in parts of the country that still supported natural vegetation, the reserves were rep-resentative of the island's diverse ecosystems. I had the opportunity to explore two of these reserves, Analamazaotra Reserve, also known as Périnet in French, and Betampun Reserve, located near the eastern port city of Toamasina (formerly Tamatave).

Being conditioned to Latin America, where a Wild West lawless-ness in the use of natural resources prevails almost everywhere, I was stunned by what I found in Madagascar. Both Malagasy reserves were guarded and respected by local people.

At Périnet, an easily accessible point on the road between the capital and the main Indian Ocean port, a dilapidated hotel from the French period catered to bird-watchers and nature seekers. Several well-maintained trails radiated into the forest, offering a choice of routes.

The biological pièce de résistance at Périnet is the indri, a teddy bear of a primate with huge, round ears, the largest surviving lemur. Indris are *fady* in Madagascar, meaning it is taboo to harm them. Their wailing territorial cries have a haunting quality that may have been likened by villagers to the voices of spirits from the other world. Indris are a given at Périnet. Several groups of them occupied the forest within easy walking distance of the hotel. Having been studied by primatologists, the animals were not shy of people and could be observed and photographed at close range. Périnet offered the makings of an ecotourism success story.

Not so at Betampun, the second reserve, which encompasses a tiny remnant of nearly extinct coastal forest. We reached the reserve only after a strenuous two-hour hike from the closest point of approach in a four-wheel-drive vehicle. Yet to my surprise, when we arrived at the boundary we found an astonishingly orderly scene. A warden and his family lived there in a small cabin provided by Eaux et Forêts. A clean and well-maintained guest house stood next to the warden's, waiting to receive visitors, who seldom arrived. We were thankful for such comfortable and welcome shelter.

Betampun Reserve had itself become an island, the only remaining patch of forest visible from a prominent hilltop. Cleared land extended right up to the reserve's boundary. However, unlike the boundary of Cerros de Amotape National Park in Perú, the boundary of Betampun was clearly and amply marked by signs in both Malagasy and French. A lack of fresh cuts and other signs of incursion indicated that the boundary was respected by local residents.

The main attraction of Betampun is the black-and-white ruffed lemur, another large and colorful primate. We could hear these animals calling in the depths of a ravine, but we were never able to see them. The forest has survived at Betampun because the terrain is so steep that no one can cultivate it. To reach the ravine's bottom, where we had heard the lemurs, it would have been necessary to rappel down. Because of the steepness of the terrain, it wasn't practical to enter the reserve. Instead, all we could do was walk along a narrow ridge that constituted one boundary and peer into the forest, hoping for a glimpse of one of Madagascar's rare birds or primates. Whereas Périnet offered ideal conditions for ecotourism, Betampun offered only disappointment. Still, the reserve was obviously being protected, a credit to Malagasy culture and the dedication of Eaux et Forêts personnel.

The survival of the Périnet and Betampun reserves might be interpreted as an encouraging sign, a hopeful indication for the future. I wish I could be so sanguine. The island's population of slash-and-burn farmers is expected to double in twenty-three years. In addition, erosion and loss of fertility of existing agricultural land are as severe as anywhere I have ever seen. Throngs of destitute peasants will have no choice but to seek new land in the remaining forests. Deforestation, conservatively estimated at 2,000 square kilometers per year in 1989, was consuming the 24,000 square kilometers of remaining evergreen forest at an appalling rate of 8 percent per year.[26] Should the rate of clearing not diminish, all remaining forest will be lost within a decade. The prospects for Madagascar and all its biological uniqueness are thus hardly better than those for West Africa.

ASIAN TROPICS

The Asian tropics are underrated and unsung in the popular perception of biodiversity. Although the large animals that roam the continent are nearly as diverse as those in Africa, many of them are poorly known except to scientists because they are so difficult to view in their forest haunts. To make matters worse, some of Asia's most distinctive fauna lives in countries such as Vietnam and Cambodia, which have been embroiled in social conflict for decades. Even where there are parks in countries at peace, many of them are difficult to gain access to and offer few or no facilities for visitors. Asian wildlife has consequently languished on the sidelines of the booming ecotourism industry.

Whereas Africa has only two rhinoceros species, Asia boasts three. Tigers, leopards, clouded leopards, snow leopards, and a host of lesser cats fill out a roster of felines unmatched by any other continent. Counterbalancing Africa's gorillas and chimps are Asia's great apes, gibbons, and orangutans. Elephants, gaurs, bantengs, tapirs, tarkins, sun bears, and sambars add to the list of impressive mammals. Yet few of these animals are readily seen in the open in broad daylight, so they are poorly known to the public. To view any of them in the wild, one must be both intrepid and persistent. Asian wildlife is not for the one-day cruise ship stopover or the quick safari-style bus tour.

Because my own experience in Asia is limited to a single three-week visit, I am unable to provide a personal perspective on the region's parks. Thus, I am obliged to draw on other sources.

Nepal

One of the brightest success stories in Asian conservation is Royal Chitwan National Park in Nepal. A royal hunting reserve since 1846, it was made a national park in 1973. Protecting a mosaic of forests and grassland, the park supports one of the most impressive herds of large mammals in Asia, including the largest population of great one-horned rhinoceroses. Tourists can view rhinos and tigers from elephant back, as did Hillary Rodham Clinton and her daughter, Chelsea, in 1996.

The history of Chitwan provides an example of how rapidly conditions around national parks can change. The Terai region, in which Chitwan is located, was largely unpopulated until the 1950s. A malaria eradication program then alleviated the primary obstacle to settlement, and migration from the Himalayan foothills began. By 1980, there were 260,000 people living in 320 villages around the park's boundary.[27] The region continues to attract migrants, and its population was growing at 6 percent annually in 1990, fast enough to double in just twelve years.

One-horned rhinoceroses are the centerpiece attraction of Chitwan. To conserve them, the Nepalese government spends nearly $2,250 per square mile on management and protection of the park, an amount that is roughly matched by tourist gate receipts.[28] Rhino poaching is kept to a minimum by the presence of 800 soldiers of the Royal Nepalese Army. Park administrators have found that offering rewards for information is the best way to thwart poaching. In 1993, for example, $3,500 was paid to informants, and thirty-seven rhino poachers were apprehended.[29] Consequently, Chitwan harbors one of the few increasing rhinoceros populations in the world.

Enforcement of park regulations against firewood gathering, cattle grazing, and other infractions is equally rigorous. In 1985, 554 violators were fined and 1,306 cattle were impounded by the park's authorities. Without constant vigilance and vigorous enforcement, requiring the presence of the Nepalese army, Chitwan's rhinos and other wildlife would be doomed. How much longer can the government sustain the will to resist such relentless pressure, a pressure that can only increase in lockstep with increases in the local population?

Thailand

The 220,000-hectare Khao Yai National Park, located about 200 kilometers northeast of Bangkok, conserves one of the largest remaining tracts of tropical moist forest in mainland Asia. Like Royal Chitwan National Park, Khao Yai is a major tourist destination, attracting 250,000 to 400,000 visitors annually.[30] For many rare species, it offers one of the last viable remnants of habitat in Thailand. But like many other parks, Khao Yai has become an island, the land surrounding it having been deforested within the past three decades.

As of 1990, about 53,000 people lived in 150 villages around the park's perimeter.[31] A majority of these people illegally occupy land classified as reserved forest, as do more than 7 million other villagers around the country. Sporadic efforts at park enforcement have generated mortal hostility between local villagers and personnel of the National Parks Division of the Royal Forest Department. Despite $400,000 in foreign funding directed toward stabilizing the economy of villagers living around the park, logging and poaching continue unabated.[32]

Pressure on the resources of Khao Yai has been exacerbated by developments elsewhere in the country. As recently as 1950, Thailand could claim more than 350,000 square kilometers of closed-canopy forest. By 1988, satellite images revealed that less than 80,000 square kilometers remained. Unusually heavy monsoon rains that year, aggravated by deforested watersheds, produced massive flooding. A stunned government reacted by banning all further logging. However, the ban perversely stimulated a tripling of the price of lumber in Bangkok as Thailand lurched from being a net exporter of forest products to being a net importer.[33]

Spurred on by higher-than-ever prices, loggers ignored the ban and redoubled their cutting efforts. Aerial surveys conducted by the Royal Forest Department in 1989, after the ban was in effect, revealed that

54 percent more forest had been logged than in the previous year.[34] High prices coupled with lax enforcement generate uncontrollable pressure on forest resources, regardless of the formal status of the land.

DOWNSIZED, DEGRADED, DEGAZETTED

Downsizing, degradation, and degazetting are steadily eroding the effective area of existing tropical parks. Because the number of new parks being created in the Tropics has dwindled to a trickle, the accelerating degradation of parklands will soon exceed the annual area of newly decreed parklands. No one knows how the two rates compare because few data exist on park degradation. It is not possible, for example, to detect the empty forest syndrome from space.

The crossover point at which parkland lost to degradation exceeds the creation of new parkland may already have been passed. Almost certainly, the point has already been passed in Africa. Asia will not be far behind. When that point is passed for the Tropics as a whole, there will be no reversing the clock, no second chance. The effort to conserve the earth's biodiversity will enter its endgame. What then?

Park degradation poses an insidious and almost intractable dilemma. Once hundreds or thousands of peasant families have helped themselves to parkland and cleared its forests for subsistence plots, the situation becomes politically delicate, even in countries with authoritarian governments. A standard political response might be as follows: "How can I deny my constituents the opportunity to grow food for their families? The park was created at another time, when conditions were different, when there was enough land for everyone. Besides, the park has benefited foreigners almost exclusively. Now the needs of our people are greater, and the park's land must help serve those needs." What response does conservation have to this riposte? If the response is to eject the squatters in order to alleviate pressure on the park, the squatters immediately become the politician's problem, a problem he or she does not want. In many countries, the option of moving people to fresh, unclaimed land elsewhere no longer exists. The frontier has closed. What then? There are no good answers.

CHAPTER 6

PROTECTING
BIODIVERSITY

OUR BELEAGUERED planet currently harbors some 9,600 bird species, 6,000 mammals, 4,000 reptiles, 4,000 amphibians, 24,000 fish, 250,000 seed plants and ferns, and countless lesser organisms.[1] Where do all these species live? The answer is, everywhere, except for the relatively tiny fraction of the earth's surface that is covered by permanent glaciers or lifeless desert. Indeed, it is the very ubiquity of biodiversity that makes the challenge of protecting it so daunting.

Much of the earth's surface has already been appropriated for purposes related to sustaining the human population. Scientists have estimated that 40 percent of the earth's net primary productivity (annual growth of plant life) is already usurped by humans, either directly as crops and fiber or indirectly as animal matter harvested from the land and water.[2] Nevertheless, the current human population of 5.8 billion is expected to double by the middle of the twenty-first century, an event that would in turn double the demands our species makes on the earth's productive capacity. If we imagine a future in which one species, *Homo sapiens,* harvests 80 percent of the earth's primary productivity for its own purposes, then the amount left over for all other species could be no more than 20 percent.

Even then, much of the remaining 20 percent would very likely be concentrated in environments little used by humans, such as rocky barrens, desert steppes, the high Arctic, and open oceans, places not distinguished by outstanding biodiversity. Instead, biodiversity tends to concentrate in precisely the environments most preferred by humans—mild, fertile, well-watered lowlands. Humans and other species are thus in competition for the same living space, a competition that will be won by humans wherever self-imposed restraints do not prevail.

The task of conserving biodiversity distills down to two basic challenges, protecting it and preserving it. Protecting biodiversity entails creating parks in accordance with a systematic plan to include within their borders most or all of the species found in a region and, ultimately, in the world as a whole. But as we saw in the case of the Middlesex Fells, protection in itself provides no guarantee of preservation, by which I mean the long-term survival of protected populations. Many conservationists regard protection and preservation as two sides of the same coin, but the two concepts are quite distinct, both in principle and in practice. Protection is a necessary first step toward preservation, the ultimate goal. But protection does not ensure preservation, as we shall see, even when the formal mechanisms of protection are airtight. To date, conservation strategists have single-mindedly emphasized protection while giving little more than a nod to preservation. In the next chapter, I will explain why preserving biodiversity is a more exacting challenge than simply conferring protection to species in formal nature preserves.

The need to protect biodiversity is obvious and basic, but designing an effective plan for protecting all of a region's biodiversity becomes a challenge burdened by difficult trade-offs. How can one protect biodiversity when it is everywhere? To do so means conserving all the earth's environments, clearly an impossibility given the burgeoning demands humans are making on natural resources worldwide. At present, the earth's nation-states have on average assigned less than 5 percent of their national territories to nature protection.[3] Such a meager allotment of the earth's land surface is not nearly enough space in which to protect all biodiversity. But the total area under protection is unlikely to grow substantially. Conservation planners thus must make the best of a bad bet. No matter what steps are taken on behalf of biological preservation, large numbers of extinctions are inevitable as humans complete their appropriation of the unprotected 95 percent of the earth's lands.

Conservation scientists would be severely challenged if asked to set aside just 5 percent of the earth's surface for nature protection. In practice, the challenge is largely a hypothetical one, for science has rarely played a decisive role in determining the size and location of protected areas. Such decisions have always resided in the realm of politics, and until recently, science was hardly a consideration.

An example from my own research will illustrate the point. Some years ago, I set out with a collaborator, Blair Winter, to investigate the status of potentially endangered bird species in Colombia and Ecuador.[4] We chose these two countries because they straddle the equator and contain some of the richest plant and animal communities on earth. Colombia is renowned among bird aficionados for harboring more avian species than any other country, more than 1,800 at latest count. Ecuador would rival Colombia for this distinction if not for its much smaller size. Both countries were then, and still are, experiencing rapid deforestation, especially in the Andes, where many of the threatened species live. And already in the 1970s, both countries had well-developed systems of parks and nature reserves, at least on paper.

Our goal was to determine how many rare bird species enjoyed some measure of security in a formally protected area. We began with a list of 164 endemic species, birds that lived in restricted portions of the two countries and nowhere else. We next had to determine precisely where each species lived, the extent of its geographic range, and its upper and lower elevational limits. Gathering such basic information for birds in North America or Europe would have posed no difficulty because legions of bird-watchers have documented the distributions of every resident species in great detail. But with the birds of Colombia and Ecuador, we stepped into an enormous void. Published accounts of the distributions of Colombian and Ecuadoran birds were so vague as to be useless for our purpose. More detailed information, such as it was, lay buried in scattered and often obscure expedition reports. Some of the species had not been recorded in decades. The only way to obtain a comprehensive picture was to examine the labels of museum specimens.

Over the next two years, Blair and I visited museums to examine all available specimens of the 164 species on our list. Altogether, we examined thousands of bird skins. Whenever a label was indecipherable or referred to a locality that could not be located on any map, we had to ignore the specimen. By pinpointing the localities where each usable specimen of each species had been collected, we prepared a distribution map for each species, drawing lines around localities to enclose the area that conservatively represented its range.

Next, using a computer, we superimposed all 164 range maps to reveal the parts of Colombia and Ecuador in which endemic bird species were most concentrated. The procedure highlighted areas where as many as twenty or more endemic species occurred together. Finally, we compared the locations of these "centers of endemism" with a map of protected areas of the two countries. We were alarmed to discover that there was little correspondence between the two maps. Few of the endemic species were protected anywhere, and all the centers of endemism except one lay far from any existing protected area.

The one exception was the Sierra Nevada de Santa Marta, a massive and isolated volcano on Colombia's Caribbean coast. Rising 5,777 meters from sea level to its glacier-clad summit, it is one of the tallest mountains on earth in the height it reaches above its base. Although twelve endemic birds occupy the mountain's forested slopes, when Sierra Nevada de Santa Marta National Park was established on part of the mountain, no one thought about birds. Ironically, the park begins 4,000 meters above sea level and encompasses the rugged, glaciated summit but includes none of the forested habitats at lower elevations where the endemic birds live. The purpose of the park was to protect scenic vistas, not biological values. And so it is with many parks around the world, including many of those in the United States.

Indeed, much of the 5 percent of the earth's land surface that is formally designated as parkland is preempted by scenic parks. Conservationists refer to this as the "rocks and ice syndrome." All over the world, serious discrepancies exist between the places protected as parks and the places that are most strategic for protecting biodiversity, as in Colombia and Ecuador. Although maximal biodiversity levels typically occur in lowland habitats, a disproportionate number of parks are located in mountainous country, which is unsuitable for agriculture. Conversely, biologically rich regions that have high agricultural potential, such as the North American tallgrass prairie, tend to be almost devoid of parks. Governments understandably strive to locate new parks in remote, unsettled areas, but few areas that offer high economic potential remain unsettled in the 1990s. Remaining wilderness is likely to be underlain by swampy, infertile, or rocky soil. A prime example is the sparsely populated Guyana Shield of northeastern South America, an enormous geologic formation that encompasses parts of Brazil, Venezuela, and the Guyanas. The white-sand soils of the Guyana Shield are notorious as some of the most barren on earth. Yet where is South America's largest park, Canaima National Park, in Venezuela, situated? Entirely on the Guyana Shield.[5] Infertile soils support slow-growing plants that strongly defend their leaves against herbivores by means of toxic chemicals. Animals in such regions as the Guyana Shield are consequently scarce, in numbers of both species and individuals. Conversely, on rich soils, where plants grow rapidly and defend their tissues less with deterrent chemicals, animal populations tend to be abundant. Unfortunately, such productive places are just the ones people want to use for agriculture. That is the bad news.

The good news is that conservation scientists have made impressive strides in devising strategies for biodiversity protection since Blair Winter and I published our findings. More crucially, governments are finally taking notice as one country after another draws up a National Biodiversity Action Plan under the auspices of the United Nations Environment Programme.[6] Another encouraging trend is the growing list of countries, both tropical and temperate, that have pledged to protect at least 10 percent of each ecosystem type represented within their borders.[7]

International conservation organizations have taken the lead in bombarding governments with suggestions and strategies for conserving biodiversity. Each organization promotes a different approach, with some placing more emphasis on quantity of biodiversity (total number of species) and others emphasizing quality of biodiversity (distinctness of species).

In the United States, the quantitative approach is exemplified by The Nature Conservancy's fifty state Natural Heritage Programs. Each state organization is responsible for compiling information about the biodiversity of that state. The status of each species is evaluated to determine whether it is secure or threatened and, if threatened, whether at the state, national, or global level. Such an approach is commendable, though working with individual species is a luxury that only a handful of the wealthiest countries can afford. In the biologically rich and economically poor countries of the Tropics, the emphasis must be directed to whole faunas or ecosystems rather than individual species.

Another organization, BirdLife International (formerly the International Council for Bird Preservation), based in Cambridge, England, compiled a book called *Putting Biodiversity on the Map,* which will help conservationists obtain more bang for the buck.[8] Following the approach Blair Winter and I used, the book defines 221 areas of avian endemism in every corner of the earth. Endemic bird areas occur on every continent and many islands, but they are overwhelmingly concentrated in the Tropics. Collectively, they account for 26 percent of all bird species in less than 5 percent of the world's land area. The book represents a call to action, highlighting as it does many important endemic bird areas that remain unprotected at present.

Conservation International (CI) has chosen to emphasize quantity of biodiversity more than quality (endemism) in its campaign for what it terms megadiversity countries. These are the countries harboring the greatest number of species within their respective regions.[9] Not surprisingly, all of CI's megadiversity countries are in the Tropics: Mexico, Colombia, Brazil, the Democratic Republic of Congo (formerly Zaire), Madagascar, and Indonesia. Although the megadiversity country approach helps focus an organization's priorities, it fails to guide planning at the subnational level. Moreover, it does not address the ultimate issue of long-term preservation, which is crucial given the political instability of some megadiversity countries.

The World Wildlife Fund, under the leadership of Eric Dinerstein, director and chief scientist of the organization's Conservation Science

Program, has adopted yet another approach. The goal is to delineate the world's ecoregions, which are defined as broad ecological formations that are internally homogeneous with respect to climate and vegetation and distinct from neighboring ecoregions.[10] Examples from the United States are the Tallgrass Prairie ecoregion of the central plains states, the Sonoran Desert of the Southwest, the Southeastern Mixed Forests, and the Eastern Cascades Forest of the Northwest. Because ecoregions are delineated on biological grounds, they often extend across national boundaries, thereby multiplying the political options for protecting them. Shared political responsibility can lead to beneficial synergisms. As one example, the environment ministers of six central African countries are meeting on a regular basis to coordinate conservation planning within the Western Congo Basin ecoregion, pooling both intellectual and financial resources in the process.

Dinerstein and his team are evaluating each ecoregion for its biodiversity value, taking into account both quantitative and qualitative criteria and assessing each one for levels of current protection and vulnerability to threats. These assessments then serve as the basis for the setting of priorities. Ecoregions that are rich in biodiversity, contain no protected areas, and are under severe threat of development receive the highest priority for conservation action. One such region is the dry forest of New Caledonia, which contains many endemic species. Already reduced to tiny, scattered remnants, the ecoregion currently lacks any formal protection.

In my view, WWF's ecoregion maps are close to optimal: neither so detailed as to be impractical as guides to action nor so sweeping as to preclude the formulation of specific recommendations. What might be viewed as a weakness by some, the lack of emphasis on individual species, is in essence a nod to reality. Quite simply, no strategy can conserve all species short of protecting all remaining undeveloped habitat everywhere, a clear impossibility. Many species will consequently fall through the cracks. This much must be recognized.

A concerted effort to protect each and every species in the world would require thousands of separate preserves, the great majority of which would necessarily be small. Conservation science has established, however, that populations of plants or animals confined to small preserves are highly vulnerable to extinction. Long-term preservation is best achieved in large protected areas. Thus, there is an inescapable trade-off between protecting species and preserving them for posterity. Given the limited resources available for nature conservation, the need to face this trade-off is critical. Conservation organizations must keep their sights clearly focused on the overriding goal of preservation, even if it means resisting urgent pleas to grant last-ditch protection to particular endangered species. In conservation no less than in business, there is an opportunity cost to every investment. It is crucial to invest wisely because extinction affords no second chance.

7

PRESERVING
BIODIVERSITY
FOR POSTERITY

NOT LONG AGO, I had the experience of visiting Hillsborough River State Park, near Tampa, Florida. The park contains a rare and beautiful example of a virgin hardwood hammock, a type of forest composed of oaks and other southern trees. At first, I was thrilled to have discovered such a fine surviving example of an increasingly rare forest type, but soon my exhilaration gave way to dismay.

The entire forest floor had been overturned, as if someone had assaulted the park with a rototiller. Not a square meter of soil remained undisturbed. Hundreds of hoofprints in the soft earth left no doubt that feral pigs were responsible for the damage. Released in more than a dozen states as an exotic game animal, feral pigs are now threatening the stability of plant communities in scores of parks around the country.[1] The damage done by these animals in Hillsborough River State Park was so extensive that a friend and I could find no oak seedlings anywhere. That beautiful forest, one of the last of its kind, had joined the living dead—an entire biological community had been re-duced to a collection of superannuated individuals that were unable to reproduce because of altered ecological conditions.

While on the subject of feral pigs, I must recount another story, an even sadder one. Great excitement spread through the ranks of bird-watchers in the mid-1970s when it was announced that a hitherto un-known species of bird had been discovered on the island of Maui, in the Hawaiian archipelago. Given the name of poouli, the attractive finchlike bird is brown above, with a white cheek and rufous ab-domen.[2] Perhaps the species was once known to ancient Polynesians, but no one in modern times had ever seen it because it lives in a small, inaccessible area on the forbiddingly steep and wet northeastern slope of Haleakala Crater. The poouli came to light during a massive effort to map and count what remained of Hawaii's highly endan-gered native bird fauna.[3] Its discovery stunned the birding community because no one had any inkling that an undescribed bird might lurk in the United States at that late date.

Now, only twenty years later, the poouli teeters on the brink of extinction, if, indeed, any of the birds still survive. Only three were found in the most recent survey of the area where the species was dis-covered, and extensive searches have failed to uncover additional indi-viduals anywhere else. Its habitat enjoys the strictest level of federal protection under the Endangered Species Act, but remoteness, inac-cessibility, and legal protection together have not prevented invasion of the poouli's habitat.

As in Hillsborough River State Park, feral pigs have devastated the forest on Maui's northern slope, devouring the fruits, seeds, and seed-lings of canopy trees and overturning the earth, so the forest cannot regenerate. I have not seen the poouli's habitat myself, but I have seen photographs of it taken before and after the invasion of feral pigs, and the stunning contrast between the two scenes left me openmouthed.

Unless the pigs can be controlled or, best of all, eliminated, the habitat of the poouli, and no doubt of other native Hawaiian birds as well, will be doomed. Yet the habitat destruction wrought by feral pigs is but one of many examples of unanticipated ecological change brought about by the invasion of exotic species. Similar accounts could fill volumes.

Ecological imbalances such as those brought about by introduced species may not always be so cataclysmic, but they almost always precipitate unwelcome change. Although such imbalances can arise in many ways, perhaps the most common and insidious way is through habitat fragmentation. Nearly everywhere, humans are pushing back the frontier, splintering and fragmenting what once were continuous habitats. As pioneers push into a new region, they naturally select the best sites, those with optimal combinations of level ground, fertile soil, availability of water, and access to transportation, schools, and other amenities. Latecomers must pick and choose from sites rejected by first comers. Steep, rocky, and barren places that are deemed unsuitable are left until last. The pattern of selection and rejection almost invariably leads to fragmentation.

As settlers spread into a region, the landscape is transmuted into a patchwork of native habitat, degraded land, pasture, cropland, and settlements, all interdigitated in a jigsaw mosaic. Whatever bits of native habitat survive are likely to be in small, disjunct patches. When such patches are all that remain of the original environment, nature begins to unravel.[4]

The initial signs that something is amiss are typically subtle, requiring a practiced professional eye. Progressive ecological deterioration and loss of species proceed so slowly that individual memories are easily deceived by the passage of time. Accurate historical records are seldom available, and change is reckoned as the absence of species that once were present. The absence of something can never be proven, so disappearances are shrugged off as artifacts of a fallible memory. And the downward slide continues.

My first awakening to the dire consequences of fragmentation came during a visit I made in early 1970 to Barro Colorado Island, a research preserve of the Smithsonian Institution in Panama. Barro Colorado Island, or BCI, as it is called by its devotees, is a 1,600-hectare (3,800-acre) island in Gatun Lake, an artificial impoundment that serves as the central portion of the Panama Canal. BCI has been protected ever since the canal was completed, just before World War I. Apart from a small clearing containing a cluster of laboratory buildings and dormitories, the island has been left to nature.

On my 1970 visit, I had the privilege of going into the forest with ornithologist Edwin Willis, who had been studying birds on the island for a decade. His special interest was a group of birds that perch over raiding swarms of army ants, not to eat the ants but to capture cockroaches, spiders, and other creatures as they flee the onslaught of the predatory ants. Dubbed "professional ant followers" by Willis, these birds exploit the ants' ability to rout any animals in their path, just as cattle egrets post themselves near the hooves of cattle, waiting for the larger animals to flush insects from the tall grass.

Willis knew exactly how many ant-following birds there were on BCI because he had been banding and observing the birds for a decade. Through close monitoring of the populations, he had noticed that the numbers of several species were declining. Two, the rufous-vented ground cuckoo and the barred woodcreeper, had disappeared entirely. A third, the ocellated antbird, had declined so radically that only a single living female remained on the island in 1970. Two years later, she disappeared, leaving two males at a reproductive dead end. During the same time span, yet another species, the bicolored antbird, declined by half. Of an original complement of six professional ant followers, only one, the spotted antbird, the smallest of the lot, had held its own.[5]

The decline or disappearance of so many ant followers prompted Willis to wonder whether other BCI birds were also in trouble. In most circumstances, such a question would have been impossible to answer because "before" data are seldom available to provide perspective on a changing environment. But BCI is an exception because it has been a haven for researchers since the 1920s.

The first to leave a record of BCI's avifauna was Frank Chapman, the distinguished chairman of the Department of Ornithology at New York's American Museum of Natural History. Author of the first field guide to North American birds, Chapman was particularly fond of BCI; for many years, he made regular visits to the island to study birds and, no doubt, to escape the slush and chill of New York winters. Chapman kept detailed notes of his observations and left a comprehensive record of the birds that inhabited the island in the early 1920s.[6]

When Willis resurveyed the birds of BCI in the early 1970s, he discovered to his great surprise that forty-five species listed by Chapman no longer occurred on the island. Many of the disappearances could be explained as the consequence of plant succession. At the time BCI was created by the rising waters of Gatun Lake, about half its area had been cleared for slash-and-burn cultivation. At that time, more than two dozen bird species that typically inhabit pastureland, brush, edges, and early second growth lived on the island in recovering areas of disturbance. As time went on, these species gradually disappeared as their habitats reverted to tall forest in a normal and predictable process. But there were other missing species, denizens of mature forest, whose disappearance could not be so easily explained. Since Chapman's time, disturbed portions of the island had matured. Why then were birds of mature forest disappearing from BCI while they remained common on the nearby mainland?

Willis published a lengthy monograph documenting what was known about the disappearances.[7] The bird species, some eighteen of them, comprised an eclectic lot that included game birds, raptors, several ant-following specialists, two wrens, and a number of others. They shared few traits except that several of them were relatively large bodied and foraged or nested near or on the ground. These weak correlations, however, were not sufficiently illuminating to suggest a chain of cause and effect, so the forces behind the extinctions remained a mystery.

In the early 1970s, when Willis published his monograph, biologists viewed extinctions as primarily chance events to which small populations were particularly susceptible. For example, an entire generation composed of individuals of the same sex could arise simply by chance in a population that was small enough, creating a reproductive dead end for the species. Furthermore, small populations are subject to inbreeding, which can lead to reduced viability and fertility. A second class of chance events pertained to environmental stresses resulting from fires, hurricanes, climate change, and other such perturbations, which could decimate a population occupying a small, circumscribed area such as that of BCI.

However, neither class of chance events provided a convincing explanation for the disappearance of BCI's birds. Common species simply dwindled year after year, as if driven into decline by some invisible hand. Although Willis found that declining species were not reproducing normally, there were few clues to explain why. Unable to reproduce fast enough to maintain themselves, population after population dwindled to zero. Science had no theory to explain these observations.

Birds are not the only organisms to have mysteriously disappeared from BCI. Several large mammals, including jaguars, pumas, and white-lipped peccaries, were observed repeatedly and even photographed in Chapman's time, but today they no longer inhabit the island. A number of reptiles also have apparently vanished or at least haven't been seen on the island in many years. One might imagine that the island carried a mysterious jinx for certain species, but scientists, rejecting superstition, searched for more mundane causes.

First to probe the mystery was the Smithsonian Institution's Eugene Morton.[8] He captured several pairs of two of the missing species, the song wren and the nightingale wren, on the nearby mainland and released them on BCI. To his satisfaction, the released birds did not desert the island. Most of the pairs remained together and soon established territories. When the breeding season arrived, the birds displayed all the appropriate behaviors, building nests and laying eggs. However, these nesting efforts failed more often than they succeeded, so in time the numbers of both species dwindled and eventually fell again to zero.

Morton's results suggested that the disappearances were not matters of chance. Some unidentified force was causing high rates of nest failure in the birds that had vanished but not in other species that continued to thrive on the island.

Bette Loiselle and William Hoppes, then graduate students at the University of Illinois, took up the challenge of solving the mystery. Perhaps, they speculated, nest predation was abnormally high on BCI. To test the idea, they deployed artificial nests of a type used by aviculturists. The wicker nests were attached to shrubs and saplings in the forest, stocked with quail eggs obtained from a local supermarket, and monitored periodically for survival of the eggs.

The results were dramatic. Sixty-nine percent of the nests set out on BCI were lost to predators, whereas only 4 percent were lost on the mainland.[9] Here at last was strong evidence of a biological force behind the extinctions of BCI birds. Still, some key questions re-mained. Why were nest predation rates so much higher on BCI, and what animals were responsible? And why were only certain birds ad-versely affected when many others were holding their own?

In science, the answer to one question can sometimes unexpect-edly shed light on another. During the early 1980s, a colleague, Louise Emmons of the Smithsonian Institution, and I conducted a study of felid predators—jaguars, pumas, and ocelots—at Cocha Cashu Bio-logical Station in Perú. To understand the ecological roles of these cats, we had also to study their prey. To estimate prey numbers, Louise spent hundreds of hours in the forest by day and by night, silently walking trails and recording her encounters with mammals of all kinds.

When we compared her data with comparably obtained data from BCI, the differences were striking.[10] Many mammals were far more abundant on BCI than at Cocha Cashu. These included both arboreal species, such as sloths and howler monkeys, and terrestrial species, such as agoutis, pacas (both large rodents), armadillos, opossums, and coati mundis (a tropical member of the raccoon family).

Why should a small island in Panama support so many more mam-mals than the Amazonian forest? The simplest explanation is the lack of top predators on BCI.[11] Jaguars, pumas, and harpy eagles disap-peared from the island decades ago, and many of the species that once constituted their prey now live in relative freedom from predators. Among the species to have profited most from the predator-free envi-ronment is the coati mundi. There are roughly twenty times as many coatis on BCI as at Cocha Cashu. Coatis, like raccoons, are oppor-tunistic nest predators and will eat eggs or baby birds whenever they find them. Thus, a superabundance of coatis could easily account for high nest predation rates on BCI. But why have only certain bird spe-cies been affected? Most likely, some species build their nests in places that are inaccessible to coatis, such as among slender vines or at the tips of supple branches high in the forest canopy.

But University of Washington ornithologist James Karr has proposed another explanation.[12] In a long-term banding study of forest birds conducted on the Panamanian mainland, he found that the annual survivorship of species that had disappeared from BCI was consistently lower than that of surviving species. Species experiencing higher mortality would have to reproduce faster in order to maintain stable populations. Such species would therefore be more vulnerable to increased nest predation than those with lower natural rates of adult mortality. Karr's findings are thus consistent with the conclusion that abnormally high rates of nest predation precipitated the disappearance of so many BCI birds.

I do not claim that the mystery of the vanished BCI birds is entirely resolved. Undoubtedly, the final story, when it is told, will involve more players and be more complicated than I have suggested. Nevertheless, it is clear that extinctions occurred on BCI because nature had fallen out of balance. There are no top predators and there are too many prey species, some of which are mesopredators—omnivores, such as the coati, that can wreak ecological havoc when abnormally abundant. Other ecological imbalances will undoubtedly emerge over the long run, but the coming changes are unpredictable, and the island has already lost a significant fraction of its birds and mammals. One can only wonder how many more species will drop out in the coming decades.

Results from BCI suggest that fragmentation of the landscape can lead to a serious form of ecological pathology. Fragmentation promptly cuts out the top predators because these animals require huge expanses of habitat in which to maintain themselves and they do not mix well with human beings. In the absence of predators, prey species increase to limits defined by the food supply. When the food supply includes baby birds or the seeds and seedlings of canopy trees, superabundant animal populations can affect the reproduction of other species, including plants as well as other animals.

After I had written several articles proposing the foregoing explanation of BCI's ills, I found myself wanting a deeper understanding of the consequences of fragmentation. I decided, however, that it would be difficult to make progress by conducting more research on BCI. Although it has lost some species, the island still harbors a rich fauna and flora, including more than 200 bird species, several dozen mammals, and more than 1,000 plants.[13] Consequently, the number of possible interactions between predators and prey, plants and dispersers, seedlings and herbivores, and so forth would number in the thousands. As an experiment in fragmentation, BCI is much too complicated to manage scientifically. Moreover, since BCI has now been isolated for eighty years, extinctions of the most vulnerable species have already occurred.

The setting I eventually selected for launching a major fragmentation project was Lago Guri, a hydroelectric impoundment in Venezuela. Like Gatun Lake in Panama, Lago Guri has inundated an entire landscape, putting all but erstwhile hilltops under water. When I first visited Lago Guri in 1990, its islands were only four years old. The impoundment is so vast that from a small boat in the middle, one cannot see the shoreline, only distant mountaintops rising above the horizon. In all parts of the lake except the very center, there are islands—hundreds of them, small and large, near and far from the mainland.

If my purpose was to study the aberrant biological processes of habitat fragments, why did I choose islands? Habitat fragments surrounded by pasture or cropland are ubiquitous in the United States and elsewhere, and they are a better analogy to the isolated parks of the future. But habitat fragments in a landscape mosaic are difficult to study because they are not fully isolated. Key animals can potentially come and go without being observed, altering results in unknown ways. Islands, on the other hand, are attractive because they afford tight control over the presence or absence of animals. Most nonflying animals are trapped when an island forms, effectively held prisoner by the surrounding lake.

One of the first tasks my collaborators and I had at Lago Guri was to census the birds of a dozen islands to generate baseline data for future reference. All the islands we selected are much smaller than BCI and were already missing most of the bird species resident on the nearby mainland. Despite reduced numbers of species, most of the islands supported high densities of birds, as many as three times the number of breeding pairs per hectare as on the mainland. The high densities could be explained in part by the fact that some species, pigeons especially, seemed to be attracted to islands as potentially predator-free nesting havens. The smaller and more isolated an island was, the more pigeons it harbored. Being strong fliers, pigeons were using the islands mainly for nesting. Most of their feeding was conducted on the mainland, to which they freely commuted.

One of the islands immediately attracted our attention. We called it Lomo, Spanish for *back*, as it is long and narrow. Most of the islands on Lago Guri bustle with bird life, but Isla Lomo stands eerily quiet. How bizarre it was to stand in the silence of Lomo's ridge top and hear a chorus of song issuing from a neighboring island only a few hundred meters away. We soon discovered that Lomo was unique among our study islands in harboring a group of capuchin monkeys. The animals had evidently become imprisoned there when water rose in the impoundment.

I was incredulous when we first encountered the monkeys. Capuchins need a lot of space. John Robinson, now director of the Wildlife Conservation Society, had studied the same capuchin species at another location in Venezuela. His principal study troop had ranged over an area encompassing nearly 200 hectares.[14] Lomo, offering only 11 hectares, seemed much too small to support a group of these wide-ranging primates. I am still mystified by the ability of the animals to persist in so small an area, but the fact that they do implies that their ecological influence on the island is some ten to twenty times greater than that of a similar group on the mainland.

Like the coatis of BCI, capuchins are opportunistic nest predators, so we immediately suspected that they might explain the absence of birds from Isla Lomo. To test this possibility, we conducted artificial nest trials, placing quail eggs on the ground and in cup nests attached to low shrubs and trees. When we returned a week later, 100 percent of the aboveground nests had been raided, as had 85 percent of the egg placements on the ground. The 15 percent of the eggs that survived on the ground were under overhanging leaves or in hollow logs, places hidden from view to an arboreal predator. Had terrestrial predators been responsible, the eggs in these sheltered spots probably would have been found. We ran similar tests on eleven additional islands, but very few nests were raided on any of them. And on the mainland, where capuchins are present at normal densities, the loss rate was a moderate 28 percent.[15]

All evidence pointed to Lomo's capuchins as the likely culprits because the island's meager fauna does not offer many other candidates. There are howler monkeys, agoutis, and iguanas, but all these are strict vegetarians. Patrolling the ground are armadillos, tortoises, lizards, and some small rodents, but most of these species do not climb trees and so could not have raided the aboveground nests. Capuchins, however, are voracious omnivores, ready to eat almost anything, including small mammals and birds as well as eggs and nestlings. I, for one, am satisfied that the onus lies on the capuchins.

Both the coatis of BCI and the capuchins of Isla Lomo are living at densities far higher than those that are normal, multiplying their ecological effects. Under normal conditions, neither coatis nor capuchins, nor even both in combination, impose prohibitive levels of nest predation. But in the abnormal circumstances of isolated and predator-free forest fragments, either species has the potential to become an agent of extinction. How many other species can also assume this doomsday role? Unfortunately, science does not yet have the answer, although hints are being uncovered by our continuing research on the Lago Guri islands.

The absence from these islands of large predators and a host of other species that perform essential ecological functions has generated a variety of bizarre but scientifically informative situations. The animal communities of the smaller islands are unlike any that would ever occur on the mainland.[16] In short, the Lago Guri islands constitute a megaexperiment aimed at revealing how ecosystems work. Specifically, our results are helping to identify the processes that maintain biodiversity in a normally functioning natural landscape.

A first-time visitor to one of our study islands would not notice anything amiss. The islands are covered by a type of tropical dry forest typical of the region. It is only when a scientist familiar with the region's flora and fauna looks at an island with a hand lens, so to speak, that aberrations become evident. One such aberration is the absence of many animals. Between 75 percent and 90 percent of the vertebrates (birds, mammals, reptiles, and amphibians) that inhabit similar forests on the mainland are missing from the smaller Lago Guri islands (those smaller than about 20 hectares, or 50 acres). Some social insects, such as honeybees and certain ants, are also missing, as are many butterflies. Yet animal species that have consistently survived on island after island have, in most cases, increased in abundance to levels far above those typical of the mainland. Among the best survivors are howler monkeys, armadillos, several small rodents, common iguanas, some smaller lizards, a tortoise species, a species of large toad, a poison arrow frog, two kinds of tarantulas, and leaf-cutter ants.[17]

No one has ever before documented such a bizarre animal community in an otherwise natural setting. Not only is the fauna composed of an extremely odd collection of species, but also from a functional standpoint it is grotesquely skewed. There are no predators of vertebrates. Nearly all seed dispersers are absent, as are some of the most important pollinators. The deficiency or lack of some ecological functions contrasts with exaggerated levels of others, especially herbivory and seed predation. Among the most abundant and conspicuous of the surviving species are folivores (herbivores that harvest leaves)—howler monkeys, iguanas, and leaf-cutter ants. On some of the islands, small rodents in great numbers emerge at night and immediately find and consume nearly every seed one puts out.

The increase in abundance of some species, particularly herbivores and seed predators, provides an explanation for the pathological ecology of these islands. In 1996, we spent several weeks camped on a 1.3-hectare island we named Iguana, as it supported a great number of these large reptiles. On our arrival, large mounds of fresh dung advised us that the island was also home to howler monkeys. All told, there proved to be 7, including 2 infants. Seven howler monkeys on 1.3 hectares of land is equivalent to 500 in a square kilometer. For comparison, howler densities on the mainland normally fall in the range of 20 to 40 per square kilometer. Isla Iguana was supporting more than fifteen times the normal density of these leaf-eating primates. In addition, it was home to a large but indeterminate number of iguanas and five large colonies of leaf-cutter ants.

Leaf-cutter ants are a distinctive denizen of the New World Tropics, often seen marching nestward bearing disk-shaped slices they have cut from the leaves of trees and other plants. Mature leaf-cutter colonies on the mainland around Lago Guri are so rare that we have yet to find one, although we know they must be there because we find dispersing queens that are seeking to found new colonies. Leaf-cutter populations explode on small islands, reaching densities of as many as five colonies per hectare, perhaps 100 times greater than the density on the mainland. Whatever regulates their numbers on the mainland is clearly absent on small islands. At the enormously elevated population levels they achieve on these islands, leaf-cutter ants are having a devastating effect on the vegetation.

Yet the effect could easily be overlooked by an untrained eye. To appreciate what is happening, it is necessary to inventory first the forest canopy and then the understory. In a healthy, self-maintaining forest, nearly every species present in the canopy will be represented in the understory as seedlings, saplings, and pole-sized juveniles. But on the islands we are studying, seedlings and saplings are scarce. The few juvenile trees in our inventories are of sufficient size to have plausibly germinated before the islands were isolated. Subsequently, ecological conditions have apparently changed drastically, such that tree reproduction has been disrupted.

The forests of the Lago Guri islands thus provide another example of the living dead. One by one, the trees that compose the existing canopy will die. Few will be replaced by others of the same kind. Most will be replaced by one of a small number of shrubs and woody vines that seem resistant to the onslaught of superabundant herbivores and seed predators.

The situation parallels one that develops in overgrazed rangelands all over the world, wherein palatable annual grasses are reduced or eliminated by overgrazing and replaced by unpalatable woody perennials. Thus, in the United States, sagebrush has spread over vast expanses of erstwhile dry grassland in the intermountain West, and prickly pear, yucca, agave, and other formidably armed plants have similarly burgeoned in parts of the Southwest.

On the Lago Guri islands, a downward ecological spiral has been unleashed by the absence of predators. Some prey species then increase explosively, with adverse consequences for other species with which they interact. Plant reproduction is particularly sensitive to distortions in the animal community. The few species that can reproduce under abnormal conditions of pollination, seed dispersal, seed predation, and herbivory increase at the expense of the many that experience reproductive failure. The eventual consequence is a gradual but profound shift in the character of the plant community, accompanied by a massive loss of species diversity. As plant species disappear, the animals that depend on them also disappear, generating an extinction spiral. In the end, all that remain are highly simplified and rudimentary ecological communities, deficient in many of the processes that maintain diversity and constructed of the most unpalatable and resilient plant species.

To be sure, the Lago Guri islands exist in a novel situation of extreme ecological distortion. One might therefore dismiss our results as exaggerated and alarmist. After all, most parks and nature preserves are much larger than the small islands we are studying. Is it reasonable to expect that similar ecological distortions will arise in larger areas in a mainland setting? I would be comforted if I thought not. But rapidly accumulating evidence speaks to the contrary.

What is happening on the Lago Guri islands is happening only slightly less dramatically in large parts of the continental United States. Years ago, settlers systematically eliminated top predators over most of the forty-eight states. Mountain lions, wolves, and grizzly bears are now confined to the wildest corners of the national territory, and direct contact with any of them is beyond the experience of 95 percent of the citizenry.

The loss of top predators had little noticeable effect on the large animal community for most of a century. During much of that time, unregulated hunting and trapping played a role similar to that of the missing predators. Following the enactment of game laws in the first half of the twentieth century, however, many mammals have surged in abundance.

Mammal species no longer kept in check by top predators have become major nuisances in large parts of the United States where hunting is restricted. Some have become road hazards (deer, moose); others browse suburban shrubbery (deer), destroy vegetable gardens (deer, woodchucks), raid trash cans (raccoons, opossums), prey on birds at feeding stations (feral house cats), impose unsupportable levels of nest predation in forest fragments (raccoons, opossums, foxes, feral house cats), and even flood suburban backyards (beavers). Overabundant mammal populations are omnipresent in much of the suburban United States.

What few people realize is that superabundant mammal populations are profoundly affecting biodiversity in the United States. For example, conservation biologist Stan Temple of the University of Wisconsin–Madison estimates that there are 2 million feral house cats in the state of Wisconsin alone and perhaps 100 million nationwide.[18] A cause-and-effect linkage between the presence of feral cats, which feed on songbirds, and declining songbird populations is strongly suggested by research conducted by Michael Soulé and his students in the canyons of San Diego County, California.[19] Canyons still inhabited by natural predators, coyotes in particular, support thriving bird communities and few house cats, but in canyons lacking coyotes, house cats abound and birds are scarce. In a counterintuitive twist, it turns out that coyotes do not prey on the nests of small birds; instead, given the opportunity, they will prey on house cats.

A superabundance of house cats and other mesopredators is not the only problem of ecological imbalance facing the United States. Just as superabundant herbivores are altering the forests of the Lago Guri islands, overbrowsing by white-tailed deer is altering the pattern of forest regeneration in much of the East. Whereas presettlement deer densities have been estimated at two to four individuals per square kilometer, contemporary deer populations have reached densities as high as forty or even sixty per square kilometer.[20] Studies conducted in Wisconsin, Massachusetts, Pennsylvania, Virginia, and Missouri agree that overabundant deer are suppressing oak and hemlock reproduction in these states.[21] A number of wildflower species are also in jeopardy. In response, The Nature Conservancy has been forced to construct fences to protect the last surviving populations of rare plant species from deer browsing. I suspect that overbrowsing by deer is responsible for many of the disappearances of plant species from the Middlesex Fells Reservation in Massachusetts. In short, ecological processes over much of the United States are out of balance, with long-term consequences that scientists are unable to predict. Few extinctions have occurred as yet, but we are clearly witnessing the signs of a gathering storm.

What is so frighteningly insidious about ecological imbalances is their subtlety. They operate in ways that are essentially invisible to the public eye. It takes a trained scientist to know that something is going wrong and, even more, to design the experiments needed to expose just what that something is. Conservation biology is still in its infancy, and its practitioners are just beginning to recognize the danger signals that foretell impending extinctions.

Lengthy time lags between cause and effect are another insidious subtlety of ecological distortions. For example, wolves and mountain lions were eliminated from the eastern United States more than a century ago, yet the consequent outbreaks of white-tailed deer, raccoons, and beavers have only recently become a matter of public or scientific attention. If the relative numbers of different species of tree seedlings on a forest floor have been distorted by abnormal levels of herbivory or seed predation, who is going to notice? The consequences may not be evident for 50 or 100 years, and by then, the changes in the forest may be irreversible.

I worry that the subtlety of ecological imbalances and the time lags required for distorted processes to run their course will provide skeptics and dissenters with arguments for inaction. The truth is that there is no time to lose in addressing these issues.

Indeed, protection becomes meaningless unless ecological processes are maintained in balance in protected areas in order to prevent a steady hemorrhaging of species. Scientists and managers can strive to minimize extinctions through interventions of many kinds, but for the long run, the only sure recipe for avoiding ecological imbalances is to retain natural areas large enough to maintain the whole interacting ecosystem, including, and especially, top predators. Of all the processes I have mentioned, the most crucial is predation. Top predators are the true indicators of ecosystem health. If they are not there, trouble is in the making.

To preserve nature intact, protected areas have to support stable populations of the most space-demanding predators. Large areas also offer the best insurance against the vicissitudes of an unforeseeable future, including global climate change. And large areas are more efficiently defended against marauding members of our own species.

That said, I do not wish to leave the impression that small reserves are of no value. Small reserves can certainly protect species of local significance, at least for the short run, though they will require greater expenditure per unit area than large reserves for management and protection.

At this point, we have a long way to go before it can be confidently said that a significant fraction of the earth's biodiversity is adequately preserved for posterity. To achieve this distant goal, we must not only get the science right but also translate the science into political and social action. Formidable challenges lie along the way. There will be opposition at every step, so we must not lack resolve.

CHAPTER *8*

TROPICAL FORESTS:
Worth More
Dead Than Alive

IF THE WORLD does not soon experience a sea change in public policy regarding tropical forests, the last tree of the primary forest will probably fall sometime before 2045. Despite the creation of new organizations to promote sustainable forestry and the continuing efforts of major international conservation organizations to promote alternatives to deforestation, all indications are that the rate of forest loss has accelerated through the 1990s.[1]

Deforestation is driven by a wide range of social and economic forces, but underlying them all is the relentless march of human population growth and the exponentially rising demand for land and forest products such growth generates. These demands are not going to slacken in the decades ahead; indeed, they will only expand.[2] Slowing down tropical deforestation, much less halting it, will therefore entail bucking powerful and inexorably growing forces. It is in this stark light that the prospects for conserving tropical forests must be considered.

Clearing of tropical forests around the world continues unabated, despite the intense attention these forests have received from conservation organizations and despite incremental progress at the political level, as demonstrated by the revised forestry policy of the World Bank.[3] Even the international conventions signed at the United Nations Conference for Environment and Development (Earth Summit) in Rio de Janeiro in 1992 have yet to achieve discernible results in reducing deforestation.[4] Is there some as yet undiscovered magic bullet that will put a halt to the ravaging of tropical forests? Analysis of the biological and economic facts leads inescapably to the conclusion that most proposed solutions to deforestation are nonstarters. Entirely new approaches are needed, especially in the area of government policy, but to date most governments have resisted the adoption of policies that would diminish the power of bureaucrats and give people a greater voice in how forest resources are managed.

Global updates on the rate of tropical deforestation have appeared at ten-year intervals, most recently in 1990. Two comprehensive sets of estimates were published that year, one by Friends of the Earth (FOE), a conservation organization, and the other by the World Resources Institute (WRI) in conjunction with the United Nations. Although both organizations employed satellite data, their estimates differ for a variety of technical reasons. FOE estimated the annual global rate of loss of tropical forests in 1989 at 142,000 square kilometers, an area equivalent to that of the state of Florida; WRI put the value at 160,000 to 200,000 square kilometers.[5]

If one conservatively takes the lower estimate and extrapolates into the future, assuming 7.5 million square kilometers of surviving forest in 1990 and no further increase in the deforestation rate, the last tree is predicted to fall in 2045.[6] Of course, the assumption of a constant rate of loss is unrealistic, given the mounting pressures on tropical forests and the land beneath them. Available evidence suggests that the deforestation rate is increasing, in which case the primary forest will vanish even before 2045.

Of course, global conditions are bound to change, and as they do, so will the assumptions on which current projections are based. Although much of the world is currently enjoying rising levels of prosperity, there are many reasons for thinking that circumstances in some tropical countries will worsen in the years ahead. Simply crossing one's fingers and hoping for miracles will not save tropical forests. Instead, substantive changes must occur in the outlooks and attitudes of governments around the world, for it is governments that will ultimately decide the fate of tropical forests.

Facing the prospect of biological Armageddon, international orga-

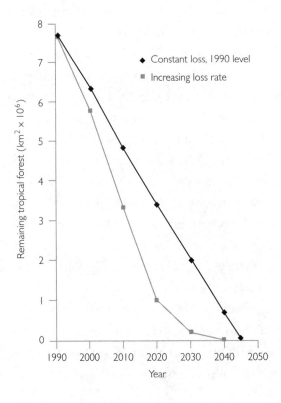

Two projections of the loss of remaining tropical evergreen forest based on surveys conducted in the late 1980s. The conservative projection assumes no change in the amount of loss per year. The steeper projection assumes that the rate of loss will continue to accelerate as it did between 1979 and 1989.

nizations of many stripes have proposed ways to slow down the rate of loss of tropical forests. Many suggestions have been offered. An often heard argument is that tropical forests should be conserved so that the biodiversity they contain can provision the world's pharmacopoeia and serve as a source of new crops, domesticated animals, and natural products of many kinds.[7] Others point to the burgeoning ecotourism industry as a force for saving tropical forests.[8] Sustainable use of forests—for example, the harvest of nontimber products such as Brazil nuts and natural rubber—has been promoted as an economically competitive alternative to timber extraction.[9] Natural forest management, a benign form of timber management that conserves diversity, is touted as a viable alternative to the continued mining of virgin forests.[10] Much as I might like to believe that these proposals offer a solution to the deforestation crisis, I am convinced that most of them represent little more than wishful thinking.

In his 1984 book *The Primary Source,* Norman Myers extols the unrealized potential of tropical forests as vast repositories of undiscovered medicines, crop plants, domesticated animals, and myriad natural products of other kinds. Much of his argument rests on the history of uses of forest products. However, in this case, I believe that the past is not a good predictor of the future. It hardly requires stating that times have changed. As a society, we have lost our rural roots. We are now participants in a highly competitive global market. To be sure, many drugs still in use are natural products, but how many of them are natural products gathered from the wild? Relatively few. Nearly every plant product for which there is a significant market is produced en masse. It makes no economic sense to do otherwise.[11]

I do not deny the prospect of undiscovered wonder drugs hiding in the depths of the world's rain forests. Merck & Company, Inc., has made a million-dollar bet on this prospect in its partnership with Costa Rica's National Biodiversity Institute (Instituto Nacional de Biodiversidad, or INBio), a private, nonprofit scientific organization dedicated to preserving the country's biodiversity. Other pharmaceutical companies are considering their own rain-forest initiatives. Nevertheless, the undeniable long-term trend is for a growing proportion of new drugs to be synthetic products. Gertrude B. Elion, George Herbert Hitchings, and Sir James W. Black shared a Nobel Prize in 1988 for inventing "designer drugs."[12] Modern molecular biology makes the future of synthetic drugs bright while dimming that of undiscovered natural products.

Even if pharmaceutical prospecting should become a major fad, which some analysts doubt, not much protected forest is needed for serious prospecting.[13] Nor does prospecting, which is a short-term activity, hold much promise for the long run. Moreover, there is no reason why existing parks and reserves could not serve the needs of prospecting. One might sympathize with Myers's vision, but in the end, his thesis fails to provide a compelling argument for increasing the total area of conserved tropical forest.

The same limitation applies to ecotourism. Said to be one of the fastest-growing industries in the world economy, ecotourism is certainly a powerful force for conservation, but it, too, has limitations. Successful promotion of an ecotourism destination requires an outstanding attraction to draw clients away from competing alternatives. For example, at Treetops Lodge in Kenya's Aberdare Mountains, the centerpiece is a salt lick visited by large mammals; at Tikal National Park in Guatemala, it is Mayan ruins; in Venezuela's Canaima National Park, it is Angel Falls, the world's highest waterfall. Wildlife viewing is sometimes the main attraction and sometimes an ancillary option, as at Tikal or Canaima. Encompassing more than 3 million hectares, Canaima is the largest park in South America and one of the largest in the world. Yet ecotourism in Canaima centers on Angel Falls, utilizing less than 10 percent of the park's area. If ecotourism is the primary motive, how can the Venezuelan government justify preservation of the rest of the park?

In a best-case scenario, ecotourism can be highly lucrative for national governments and entrepreneurs, though less often for local residents and indigenous peoples. The outstanding example of successful ecotourism is that of Rwanda's mountain gorillas. When social conditions in that troubled country permit, tourists can commune with gorillas at close range for a fee of $170. In 1988 (before the civil war), gorilla viewing at Parc National des Volcans contributed $1 million to Rwanda's economy, and indirectly it may have contributed as much as $9 million, making it one of the country's top foreign exchange earners.[14] Under favorable circumstances such as these, ecotourism can justify habitat preservation, but such circumstances are relatively rare. More typical are the hundreds of parks that are hardly known, either to citizens or to foreigners. How are these to be economically justified through ecotourism?

A brief digression underscores my point. In 1985 I spent a month in Ecuador researching migratory birds. A need to rent vehicles and to make other travel arrangements brought me into Ecuador's two main cities, Guayaquil and Quito. In both cities, I conducted a minisurvey to satisfy a matter of curiosity.

I walked into several travel agencies at random and made a simple request: I asked for information about Ecuadoran national parks. The most frequent response was a blank stare. "We don't have any information about national parks," the uncomprehending attendant would reply. Several times, the person on duty excused himself or herself to relay the question to someone in a hidden back office, but the answer was the same. In two or three of the agencies, the question elicited a different response. "Oh! You mean the Galápagos. Yes, we have information about the Galápagos." "No," I would explain, "I would like information about Ecuador's *other* national parks. There are several of them." Again the uncomprehending stare. In not one of the seven or eight agencies I polled was anyone even aware that Ecuador had five national parks in addition to Galápagos National Park.[15]

One tropical country in which parks have been accorded a high place on the official agenda is Costa Rica, a country endowed with the rarest of blessings, an enlightened government. Having few natural resources or other economic assets, Costa Rica has made parks the centerpiece of its substantial tourism industry. Alluring photographs of parks and their wildlife embellish the walls of every travel agency. U.S. citizens flock to Costa Rica by the thousands to visit parks and often little else. But is Costa Rica a good model for the rest of the Tropics?

Not every country can develop ecotourism as Costa Rica has, for many reasons. For one, tourists from the developed world are reluctant to visit many tropical countries. Over large parts of the tropical world, the residents speak an incomprehensible language, services are deficient, hygiene is lax, crime is rampant, and beggars assault foreigners on every downtown street corner of the capital. Heavily armed soldiers are conspicuous in the cities and halt traffic at frequent checkpoints along rural roads. It is little wonder that so many of the world's rain-forest parks have not been developed for tourism. With social conditions such as these, they are going to stay that way. Ecotourism under such circumstances appeals only to an adventurous few.

Ecotourism in the rain forest has other drawbacks. First-time visitors are often awed by the rain forest, simultaneously attracted by the cool, deep shade and giant trees and repelled by the fear of myriad imagined dangers. In the absence of positive stimuli, the novelty soon wears off. Tourists want to see animals, but animals are notoriously hard to view in the dense, multitiered foliage. Restricted visibility means that most animals are not detected until the visitor is already

well within the animal's flight distance, the distance at which a crea-
ture flees in the presence of a human. Tourists often come away dis-
appointed.

Poor visibility in the forest and skittish birds and mammals pose an
unresolvable dilemma for ecotourism guides. Trails in the forest are
typically narrow, so people must progress in single file. Normally, the
guide goes first, and it is the guide who spots the toucan or the pec-
cary. If the quarry doesn't bolt that instant, the guide may be able to
point it out to the second person in line. The third will be extremely
lucky to have a glimpse. The fourth in line might as well not be there.
Best results are obtained when one is alone, but a lone novice would
not have the trained senses needed to detect, much less to view, the
forest's wildlife.

The difficulty of observing wildlife in the tropical forest is an intractable problem for ecotourism. I recently met a longtime ecotourism guide, who told me he was quitting the job for precisely this
reason. It was simply beyond his power to ensure client satisfaction.
Too many customers went away complaining: the humidity was un-
comfortable, insects assaulted them, and the animals they had found
so appealing on the television screen were nowhere to be seen.

Rain-forest parks that do not boast special attractions—a salt lick, a
beach, world-class ruins, a waterfall—are going to be a hard sell. They
will attract a steady but scant flow of connoisseurs and devotees, those
who have the skill and patience to discover the forest's animal inhabitants, but the volume of business is unlikely to provide strong economic justification to a government that must employ guards to keep
poachers and squatters out. With so many undeveloped rain-forest
parks in the world (there are literally hundreds of them), how can the
promise of ecotourism persuade governments to set aside more? It is
a tough question.

Another gambit of international conservation, one with the potential to justify the retention of natural tropical forests over large areas, is sustainable use. In the current fervor for sustainable development, the term has an appealing, friendly ring. But what is sustainable use, and is it really friendly to biodiversity? More to the point, is it economically viable?

In principle, sustainable use is any economic activity that is not self-limiting, such that current practices do not diminish the potential for future yields. For example, sustainable forestry has come to mean a combination of logging and replanting so that the forest continues to regenerate despite periodic harvests. Whether or not sustainable use is friendly to biodiversity depends on how it is implemented. If profits are the primary goal, then "sustainable use" is unlikely to be practiced in a manner friendly to biodiversity, for there is an inevitable trade-off between intensity of use and retention of biodiversity.

But even strategies that are friendly to biodiversity have their limits. Two approaches that are fashionable at the moment highlight the problems. One, the harvest of so-called nontimber forest products such as nuts, is the more benign approach because it typically entails little or no hands-on management.[16] The other, known as natural forest management, is based on timber harvest and involves widely varied levels of intervention, depending on specific management objectives.[17]

The harvesting and marketing of nontimber forest products, which can be fruits, nuts, gums, resins, animals, or anything else produced by forests that doesn't qualify as timber, is a traditional practice of local people nearly everywhere forests are found. Nontimber products can provide supplemental income to subsistence dwellers even in the United States. In the New Jersey Pine Barrens and in the pinelands of the South, for example, there are people who rake and gather pine straw to be baled as mulch. These are people on the lowest economic rung, as is true of all the people I have met in other parts of the world

who gather nontimber forest products. In the absence of mechanization or other means of enhancing productivity, output per person is
necessarily low.

Several commodities of significance to international trade are
largely or entirely harvested from natural tropical forests, among them
Brazil nuts, allspice, chicle, and rattan. Brazil nuts and natural rubber
are two of the best-known and economically most important nontimber forest products. In the Brazilian Amazon, rural people called *seringueiros* have harvested these commodities for more than a century. By
interviewing a number of *seringueiros* in the state of Acre, anthropologist Stephan Schwartzman found that the average family gathered
750 kilograms of natural rubber and 4,500 kilograms of Brazil nuts
per year from 200 hectares of forest.[18] At prices prevailing in 1989, the
harvest yielded an income equivalent to U.S.$960. For a family of four,
that amounts to $240 per person, but because many *seringueiro* fami-
lies are large, the per capita earnings are typically less.

On an area basis, the economic return was a mere $4.80 per hectare
per year. Almost any other kind of land use, including slash-and-burn
agriculture, easily achieves a higher return per unit area. Herein lies
the Achilles' heel of the harvest of nontimber forest products as a
form of sustainable use. It is an extensive form of land use in a world
in which economic pressures demand ever more intensive forms of
use. The hand gathering of forest products is rapidly becoming an
anachronism, except in remote, roadless areas where lack of transportation forecloses other opportunities.

Chico Mendes, the leader of the *seringueiros'* union, was murdered
in 1988 because his efforts to preserve the lifestyle of the *seringueiros*
of Acre stood in the way of powerful landowners bent on converting
the forest to cattle pasture.[19] In the absence of well-defined and legally
enforced land rights, the clash of interests led inevitably to violence.
But Mendes's murder at least precipitated protective measures for the
seringueiros, though their future is by no means certain.

Embarrassed by the glare of international publicity that followed Mendes's assassination, the Brazilian government formally established *reservas extrativistas* (extractive reserves) amounting to 19,300 square kilometers, or about 13 percent of the state of Acre.[20] Another 25,500 square kilometers of extractive reserves were set aside in other Amazonian states. At that point, things began to look better for the rubber tappers. Thanks to the international furor over the death of Chico Mendes, the government was forced to acknowledge that the *seringueiros* had some rights.

Ironically, during the same period, other international forces were inadvertently working against the *seringueiros*. Responding to a public outcry over accelerating rates of deforestation in the Amazon region and alarmed by the vast amount of carbon dioxide being released into the atmosphere by burning of the forest, the U.S. government began to pressure the Brazilian government to reduce the rate of deforestation. Several prominent economists believed that much of the deforestation was being fomented by subsidy programs designed to stimulate the expansion of cattle ranching in the Amazon.[21] Get rid of the subsidies, the argument went, and the rate of land clearing would decrease.

At the same time, the Brazilian government wanted to reduce subsidies for other reasons. Brazil had joined the ranks of industrialized nations and become a major exporter of such high-technology goods as arms and aircraft. In order to expand its role as a player in the international economy, Brazil needed to comply with the requirements of GATT (the General Agreement on Tariffs and Trade) by cutting subsidies on domestically produced goods. Among the subsidies slashed was one that had supported a high price for natural rubber from the Amazon. Rubber tapping, a marginal lifestyle at best, ceased to be an economically viable proposition in an open market.

A Brazilian friend of mine, while traveling into the remote headwaters of the Río Juruá in 1994, passed one overgrown homestead after another. Economically defeated rubber tappers and their families had abandoned a region that offered no other economic opportunities. In search of work, they were leaving the forest and moving into the burgeoning cities of the Amazon, where they added to the squalid shantytowns that form the urban perimeter.

Without rubber tappers to provide an economic justification, what will happen to the huge *reservas extrativistas?* The legal status of the land is no more permanent than the subsidies that led to the creation of the reserves in the first place. The next economic wave to sweep the region will rewrite the story.

The outlook for other nontimber forest products is equally precarious, but more for biological and economic reasons than for political ones. Two examples illustrate my point.

Let me begin with a biological example. In the summer of 1995, I had occasion to travel up two rivers in the Peruvian Amazon, the Río Pariamanu and the Río de las Piedras. For half a century, the two rivers have represented the epicenter of Perú's Brazil nut industry. Brazil nuts were the region's only significant product until the construction of a road and the discovery of gold irrevocably altered the local economic equation, eclipsing Brazil nuts. Much of the land nevertheless remains in Brazil nut concessions licensed by the government.

This is the only part of Perú where Brazil nut trees reach economically viable densities of two to three trees per hectare. Many of the trees are magnificent specimens, soaring to fifty meters or more above the forest floor atop trunks a meter and a half in diameter. The ground beneath is littered with the cracked husks of the last harvest season. The superficial impression is that all is well in the Brazil nut concessions. But a closer look reveals otherwise. Although the mature trees are alive and well, there is not an immature tree to be found anywhere. The systematic collecting of nuts for decade upon decade has left no seeds for reproduction. The still-producing adult trees are approaching senescence. When they die, Brazil nuts will become just one more Amazonian boom-and-bust story.

My second example, which is economic, illustrates how innovation can render a traditional product obsolete. I have a boyhood memory of passing by miles and miles of longleaf pines in southern Georgia. Each tree was scarred by slashes and carried a small tin can for collecting resin. The resin would be distilled into turpentine, which was at that time the leading paint thinner. Not long afterward, the turpentine industry crashed when less expensive petroleum products took over the paint thinner market. The longleaf pines, their value as resin sources diminished, were systematically harvested for timber.

Many nontimber products that were once gathered from natural forests are now mass-produced in cultivation. Few of us today are aware that cola nuts, palm hearts, chocolate, the plants that yield rotenone (an insecticide frequently used in home gardens), and other commonplace commodities were once harvested by hand from tropical forests. The few source plants that have not been brought into cultivation are those that require decades to mature, such as the trees that produce Brazil nuts and chicle. Were that not the case, both products would undoubtedly be produced on plantations, as essentially all the world's natural rubber now is.

In short, sustainable use of nontimber products is an admirable notion, but it is too burdened with fragile assumptions to be the basis for a long-term conservation plan. Too many plant products and all animal populations are subject to overexploitation, especially when there is open access to the resource. Moreover, nontimber products by themselves rarely generate enough economic return to justify the conservation of large areas of tropical forest.

One option is to step up the scale of intensification by implementing a multiple-use approach known as natural forest management.

Here, the main emphasis shifts to timber production, but other uses, including hunting, recreation, and harvest of nontimber products, are allowed.

Natural forest management differs from simple timber removal in that sustainability is an explicit goal. There are few other guidelines except a commitment to retain a "natural" forest of native species that is not strictly a plantation. The level of exploitation can range from light to heavy, and many forms of intervention are practiced, including cutting of vines and lianas, poisoning or girdling of trees that have no commercial value, and planting of seeds or seedlings of desirable species. Such silvicultural treatments clearly reduce diversity, but they do not extinguish it altogether. Natural forest management is thus biologically preferable to more intensive forms of land use.

Natural forest management has a long history in the Tropics. The British, French, and Dutch colonial governments all had vigorous forestry programs in their overseas empires. Colonial governments recognized the immense potential of tropical forests for producing wealth and took seriously the job of managing them, imagining themselves in for the long haul. Some of the most distinguished botanists and foresters of the era worked for or with colonial forestry departments, applying great ingenuity to the challenge of maximizing returns from managed natural forests.

Superficial appearances to the contrary, no two forests are the same with respect to species composition, growth requirements of commercial species, competition from vines and noncommercial species, and other factors relevant to management. Management therefore has to be custom tailored to the conditions of each locale. Thus, instituting a natural forest management plan is far from easy and the challenges are many, as illustrated by the following four cases.

One of the earliest efforts in tropical forest management was initiated by the British in the former colony of Malaysia. Malaysian lowland forests were some of the most valuable in the world, being heavily stocked with towering, columnar-trunked meranti and other dipterocarp trees. These forests were also home to a great diversity of tree species, a fact that challenges managers to encourage the regeneration of high-value species while discouraging the growth of others.

The management strategy adopted by British and Malaysian foresters rested on a peculiarity of Southeast Asian forests, the so-called masting habit. The term *masting* refers to mass fruiting episodes that involve many species at once and occur at irregular, multiyear intervals. Dipterocarp seedlings appear only in the wake of masting events, at which time the forest floor may be carpeted with them. Forest managers were encouraged to wait until seedlings were present before harvesting mature trees. In principle, established seedlings would then regenerate a second crop of dipterocarps in seventy years, and the cycle would continue indefinitely.[22] But it proved impractical to suspend an entire industry between masting events, so mature trees were cut without regard to the presence of dipterocarp seedlings.

In West Africa, British foresters and their African counterparts encountered a set of limitations quite different from those in Malaysia. Here, densities of the desired species, mostly so-called African mahoganies, were low, frequently only one or two trees per hectare. Seedlings were correspondingly scarce.[23] To increase densities, seedlings had to be produced in nurseries and transplanted to the forest or to prepared openings in the forest. But African mahoganies grow slowly in comparison with other tree species, so the saplings tend to be overwhelmed by competitors. Efforts at commercial management of African mahoganies consequently proved to be labor-intensive and economically unrewarding.

Natural forest management did not enter the American Tropics until the 1980s, when foresters from the University of Wageningen, Netherlands, initiated a major project in Suriname.[24] A key innovation was the decision to manage for a wide range of species, as many as fifty, instead of merely a select few. The foresters reasoned that harvesting more species would allow them to utilize a larger fraction of the wood in the forest. More species also meant more variability in biological properties. Whereas African mahoganies had to be started from seed because saplings were too scarce to be relied on, many useful species in Suriname were well represented by saplings. Thus, undersized trees already present could grow up to replace those harvested, thereby greatly shortening the harvest cycle. In fact, the Dutch management system produced as much as a fourfold increase in the growth of commercial tree species. Other efforts to enhance the economic performance of natural tropical forests have not been nearly as successful.

Another successful effort in tropical forest management was orchestrated in the 1990s by a U.S. forester, Gary Hartshorn (now executive director of the Organization for Tropical Studies at Duke University), and implemented by the Amuesha indigenous community in Perú's Palcazu valley with the support of USAID. Hartshorn made the maintenance of biodiversity an explicit management objective. In Hartshorn's method, the area to be managed is surveyed and a harvest schedule is planned for a series of narrow strips up to 500 meters long and 30 meters wide. Then, in principle, about 3 percent of the strips are harvested each year to allow for a thirty-year rotation time. When harvested, a strip is entirely cleared. The cut is assigned to various uses, according to species: sawtimber, roundwood (posts and poles), specialty products, and charcoal. Everything is used. Harvested strips are allowed to regenerate from natural seed sources in the adjacent forest. Preliminary indications are that the regrowth is as rich in species as the primary forest. By confining the harvest to strips, a large volume of wood and wood products can be obtained from a relatively small area, thereby, in theory, relieving pressure on the remaining primary forest.[25]

Now for the punch line. Not one of these management systems proved viable: all of them faltered on one or more grounds.

The Malaysian system passed into historical irrelevance when virtually the entire lowland forest estate of the Malay Peninsula was converted to oil palm and rubber plantations. Quite simply, plantations are a more remunerative use of the land. I have driven myself from the Cameron Highlands, north of Kuala Lumpur, to Singapore, at the southern tip of the peninsula: hardly a hectare of natural lowland forest remains along the way.

Efforts to manage African mahoganies faltered on technical grounds. The restocking and subsequent regrowth of commercial species was disappointing, and the system was abandoned in the mid-1960s. But even without technical deficiencies, managed forests soon would have yielded to demographic pressure. In the words of Chelunor Nwoboshi, a West African forester, "Under the prevailing circumstances, forestry needs evidence of impressive socioeconomic returns from forest lands to ward off the competition from other land uses. For example, only recently Nigeria lost about 10,000 hectares of forests in the Okomu Reserve—one of the centers of natural forest management—to oil palm cultivation, and an estimated additional 280,000 hectares of productive forests were lost throughout the country."[26]

In Suriname, a rebellion against the dictatorship of Dési Bouterse created danger and uncertainty in the interior of the country, forcing the Wageningen research team to withdraw. Similarly, when Perú's Palcazu valley was invaded by Sendero Luminoso guerrillas, all USAID personnel were required to evacuate.

As these examples poignantly illustrate, forest management can be successfully practiced only under the most stable social and economic conditions. Forestry policy must carry forward from one government to the next. Continuity of land use must be ensured. Prompt and effective enforcement of land rights must be institutionalized to deter illegal activities, such as invasion of forestland by squatters. And the possibility of dictatorship and armed insurrection must be as close to zero as it reasonably can be. If these conditions cannot be met with confidence, what corporation or government will make a thirty- or seventy-year bet on the future?

The answer is obvious. But even if there were grounds for confidence in the prospects for long-term social and political stability, I would remain pessimistic about the potential for natural forest management in the Tropics. It is simply not economically competitive with alternative land uses. Consider the burden of limitations that must be overcome.

Most of the space in tropical forests is occupied by hundreds of species of trees that have little or no market value. Growth rates are low because the crowns of most trees reside in the shade. The efficiency with which desired species can be produced is therefore extremely low.

Moreover, with the exception of a few trees such as mahoganies, tropical woods are not highly desired. Because they are unusually dense, the wood grows slowly, damages saws, is expensive to transport, and, except for special uses, does not bring a premium price.[27]

And, as discussed earlier, management is costly and fraught with pitfalls. The biology of many commercial species is essentially unknown. Management practices appropriate to each type of forest must be designed de novo, and mistakes may not become evident for years. Even when a management plan is devised by experienced foresters, satisfactory results are not ensured, as demonstrated by the case of the African mahoganies.

Last but not least, there is the bugaboo of opportunity cost. Given the social, political, and economic uncertainties inherent in developing countries, investors require exceptionally high rates of return, typically in excess of 10 percent per annum. No forest management project can expect to bring in such a high rate of return or anything close to it.

Reluctantly, I have to conclude that natural forest management in the Tropics has no chance of being practiced for its own merits. Quite simply, natural forests cannot compete with alternative land uses, specifically, plantation forestry and agriculture. Any country that commits to natural forest management will do so not for economic reasons but for political reasons, perhaps as part of a comprehensive land-use plan.

Even if natural forest management were to be widely practiced in the Tropics, biodiversity would inevitably suffer. Management means intervention, and often the interventions are of a drastic kind: clearing away of the understory, cutting of all lianas, poisoning of non-commercial species, removal of all or most of the canopy at harvest. No one knows what these treatments would do to either plant or animal diversity if carried out over two or three harvest cycles.

Certainly, the more intensive the management, the greater the consequences for biodiversity. If hunting is allowed, animal populations will suffer, risking ecological imbalances that could cause reproductive failure in important tree species and trigger extinction spirals. At best, natural forest management would be a mixed blessing. At worst, it would be just one more means of accelerating extinction.

Herein lies the fundamental weakness of the sustainable use concept as it relates to biodiversity. Economic pressures will always be in the direction of intensifying use, with less tolerance for noncommercial species and a greater emphasis on more intensive management practices and shorter rotation cycles. At what point does one say, "Stop! Enough!"? Sustainable use admits of no line in the sand. The burden of proof will always be on proponents of conservation to demonstrate that a given management plan will jeopardize biodiversity. That was true in the controversy over the northern spotted owl in the Pacific Northwest of the United States.[28] It will be true in other cases in the future.

In short, sustainable use will do little to diminish the biodiversity crisis. If anything, it may help fuel the crisis by creating complacency about the future of tropical forests. When conservation organizations begin to advocate sustainable use of tropical forests, it is a signal that conservation is on the run. Starting down the slippery road to sustainable use is stepping back from that crucial line in the sand that defines one's beliefs and principles. Sustainable use represents a gray zone where politics, economics, and social pressures, not science, decide what is good for humans, with scarcely a nod to nature. For ignoring these realities, the conservation community risks a rude awakening.

Whether we like it or not, tropical forests are worth more dead than alive. Nothing can save them short of a sea change in public opinion that registers not only in politicians' statements but also in their actions. Saving biodiversity will have to become a global obsession, not merely a pastime.

CHAPTER 9

FROM WILDLANDS TO WASTELAND:
Land Use and the Mirage of Sustainable Development

SUSTAINABLE DEVELOPMENT has become the mantra of the conservation movement, appealing to everyone, business interests and conservationists alike. Why? Because like apple pie and motherhood, no one can be against it. That is because the term is seldom rigorously defined. In the absence of a clear definition, sustainable development means anything anyone wants it to mean.[1]

Ambiguity of definition has spawned a vast literature purporting to offer pathways to the nirvana of sustainable development. The proliferation of formulaic solutions, many of them contradictory, has understandably led to confusion and a creeping ennui and disillusionment with the whole concept. But loss of momentum toward sustainable development would be tragic because sustainable development is not a luxury. It is a necessity if there is to be peace and prosperity in the world of the future. The alternative is exhaustion of natural resources, crushing poverty, and social anarchy.[2]

What, then, is sustainable development? To a scientist, the definition is easy. Development is sustainable when outputs are balanced in kind by inputs.[3] An agricultural system, for example, would be sustainable if the outputs, the nutrients contained in the crop and soil that are lost to erosion, were balanced by inputs, in the form of fertilizer and new soil created by weathering of the underlying bedrock. All additional inputs, such as energy, water, chemicals, and fertilizer, would also have to be sustainably produced. Chemicals and fertilizers would have to be elaborated from renewable materials, using renewable sources of energy. Buildup or runoff of artificial chemicals would not exceed the assimilative capacity of the environment. Energy consumed in crop production, such as that used to power tractors and irrigation pumps, would have to come from renewable sources such as solar collection or hydroelectric generation. Any reliance on fossil fuels and fossil water, both nonrenewable resources, would, of course, be precluded.

When vague platitudes are cast aside and a technically precise definition is enunciated, the truth about sustainable development becomes painfully clear. Given the expanding human population, the competitive nature of the global economy, and our collective obsession with maximizing economic growth, sustainable development is currently unattainable. Certainly, sustainable development is not being practiced by any society that lives in the modern money economy. Indeed, it cannot be achieved by the world as a whole without structural adjustments so radical as to be inconceivable to governments and citizens alike. The current unattainability of sustainable development, however, should not discourage reforms that make incremental steps toward it. But let us not kid ourselves: no national economy will achieve sustainable development for a long time.

We should not kid ourselves about another point as well. Sustainable development has been touted as the salvation of biodiversity. True, biodiversity will not long survive without significant progress toward sustainable development, but to equate sustainable development with the perpetuation of biodiversity is a mistake. Sustainable development means development, just as surely as sustainable use means the exploitation of natural resources. Although there would be caveats and restrictions under sustainable development, the primary goal would still be the production of goods and services for human beings. There is nothing about sustainable development that requires the existence of biodiversity or wild nature. Those are separate issues, to be debated and decided on other grounds.

We are living through a transitional period marked by ideological countercurrents. Sustainable development has become the rallying cry of such organizations as the World Bank and USAID even as the global economy recedes ever further from sustainability in its inexorable push toward greater intensification of use of all renewable natural resources. Only rarely are renewable resources managed sustainably. More typically, they are treated, de facto, as if they were nonrenewable. One sees this in the conversion of forestland to other uses, the systematic overgrazing of grasslands, the exhaustion of fisheries, the profligate abuse of soil, and the depletion of aquifers. The litany is depressingly familiar.

Still, forests and grasslands will regrow if left to rest, fisheries will recover if their ecosystems have not been fundamentally altered, and aquifers will recharge. The one renewable resource to which intensification does irreparable damage is soil. Soil I regard as humanity's most precious asset. Everything depends on soil: agriculture, forests, grasslands, nature itself. Like air and water, it has no substitute.

Whereas intensification of agricultural production extracts a price from the soil, traditional slash-and-burn agriculture causes little or no soil erosion. To be sure, there may be temporary but recoverable losses of nutrients, but erosion is minimal on level ground because the soil is never overturned and exposed to the elements.[4] Unsustainable erosion goes hand in hand with mechanization.

Erosion has become a serious form of environmental degradation in the Tropics and elsewhere as the introduction of mechanization, fertilizers, and methodology of the green revolution have replaced slash-and-burn agriculture with intensive cultivation, and as the number of cropping cycles per annum has increased from one to two and, in some areas, from two to three. Bangladesh, for example, could not feed itself if farmers there did not harvest three rice crops per year.

Unacceptable rates of soil erosion are a feature of modern mechanized agriculture worldwide. Half the topsoil of the midwestern farm belt of the United States is thought to have washed into the Gulf of Mexico since the land was first plowed a century and a half ago. A typical farm in Missouri loses twenty tons of soil per hectare per year, whereas the rate of new soil formation is only one ton per hectare per year.[5] The U.S. Department of Agriculture considers an annual loss of five tons per hectare to be acceptable when clearly, for the long run, it is not.[6] A rate that is truly sustainable simply cannot be attained with current agricultural methods. Rather than confront reality by prescribing measures that would be politically unpopular, the Department of Agriculture plays make-believe. In short, the whole world is doing the same with its renewable natural resources.

Lying at the heart of sustainable development is stabilization of land use. When land use changes, as when a forest is cleared to make way for pasture or cropland, the original biodiversity is eliminated wholesale. More subtle, but equally devastating to native grasslands, is "pasture improvement," a euphemism for substituting exotic for native grasses. Native biota can be maintained only when the natural vegetation cover is preserved, that is, when forest is retained as forest and when grassland is pastured without overgrazing, plowing, or introduction of exotic forage species. Once humans have altered the natural vegetation cover, recovery of the original plant and animal community via the process of succession can require centuries. Attempts to accelerate succession, such as by seeding native plant species, have achieved limited success.[7] For the sake of biodiversity, therefore, the natural vegetation cover must be retained to the greatest extent practicable.

Scientists have developed criteria for optimizing land use based on evaluations of such factors as climate, soils, topography, and access to water. Application of the relevant criteria can lead to definition of the "best" use or range of appropriate uses for a given element of the landscape. Steep terrain, for example, is vulnerable to erosion and is therefore inappropriate for cultivation; pasture or tree plantations would be more appropriate. Land that is too dry to support agriculture should not be cropped if available water supplies contain salts or must be pumped unsustainably from underground aquifers. These examples are simple and self-evident.

Land-use decisions become more sophisticated when economic criteria, such as return on investment, are combined with purely physical criteria. Herein lies the danger to biodiversity because economic conditions are rarely stable over long periods, so changes in land use are an ever present possibility. Historically, instability has been the rule.[8] At the time of the American Civil War (1861–1865), nearly the entire Piedmont region of the eastern United States lacked forest cover, to the degree that encamped soldiers were often hard pressed to obtain fuelwood. Exhaustion of the soil and the availability of rich, new land in the Midwest prompted large-scale abandonment of Piedmont lands after the war ended. Depressed prices of farm products during the Great Depression of the 1930s forced a second round of land abandonment. Since World War II, rising levels of agricultural productivity have increased competition among farmers, causing many to sell out to larger, more efficient growers or to convert their land to other uses, such as plantation forestry or urban expansion. Economic forces have thus impelled the consolidation of intensive agriculture on the most appropriate land. Consequently, much of the Piedmont that was cultivated in the nineteenth and early twentieth centuries is currently forested, as it was originally, but the cost in lost biodiversity has been substantial.

Land-use patterns in Europe and North America are relatively stable compared with those in many developing countries. Stability has come with the closing of the frontier and the regulation of agricultural markets through a wide range of price support programs.[9] Land of secondary quality is employed for pasture or forestry. There is relatively little wasteland, here defined as once productive land that has been abandoned in such a deteriorated state that it has no potential use under current economic and technological constraints.

Economic development follows a different course in many tropical countries, where societies are polarized between haves and have-nots. The urban-based haves control the economy and the politics. The have-nots are largely left to fend for themselves. Lacking regular employment and enjoying few social benefits, they manage barely to survive by selling trinkets on the sidewalks of major cities or practicing subsistence agriculture in the countryside. Orderly processes, if any can be said to exist, are mainly a feature of the cities and towns. Land use in the countryside can be an unregulated free-for-all, especially where there is still a frontier.

On the frontier, road construction marks the first stage of a downhill land-use cascade. Segments of the Transamazonian Highway system are etched in fire in satellite images taken at night in the dry season, when newly cleared land is burned. More than 6,000 fires have been counted in a single scene 180 kilometers on a side. In the frenzied first-come-first-served ethic of the frontier, land is settled regardless of its quality or suitability for sustained use. Competitive economics do not matter because subsistence, by definition, lies outside the money economy. Subsistence farmers are not competing; they are surviving.

The economic desperation that drives an estimated 500 million people to practice subsistence agriculture in tropical forests poses an extreme threat to the world's remaining tropical wildlands.[10] In a competitive agricultural system, there are incentives for enlightened land stewardship because profits derive directly from the productivity of the land. Land that is inherently of poor quality fails to yield enough to cover the costs of production and is allocated to less intensive uses.

But in a subsistence system, all effort is directed toward survival. There is no profit for reinvestment in the future. Seldom is there cash on hand, so fertilizers are rarely employed, even when they could boost yields significantly. For a family trapped in abject poverty, there is no secure tomorrow. There are only uncertainties. In the face of uncertainty, opportunism reigns. So long as wildlands remain anywhere, they will be seen as opportunities by people who have no other alternatives.

Millions of slash-and-burn subsistence farmers, abetted by loggers and cattle ranchers, constitute the principal driving force behind the land-use cascade, the process by which wildlands are appropriated for human use and then exploited until exhaustion brings about abandonment.[11] Overall, the world is experiencing a net loss of wildlands and a net gain of wasteland, a situation that cannot long continue if the goal of attaining sustainable development is to be taken seriously.

The accumulation of wasteland is an indication of the unsustainability of human activities. Since 1950, an area worldwide equivalent to the combined national territories of China and India has been abandoned to wasteland.[12] If one also counts as wasteland those areas

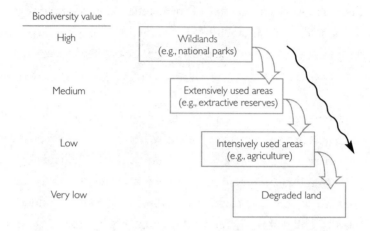

Biodiversity value

High — Wildlands (e.g., national parks)

Medium — Extensively used areas (e.g., extractive reserves)

Low — Intensively used areas (e.g., agriculture)

Very low — Degraded land

The land-use cascade. Restoration efforts will be needed to reverse the arrows.

taken over since 1950 by industry, urban expansion, and road building, the total is substantially larger.[13]

In a world faced with the necessity of feeding an expanding human population, the net conversion of wildland to wasteland can hardly be seen as rational. Here, as in other cases of natural resource use, an inherent conflict arises between the interests of individuals and the interests of society as a whole. Whereas it is perfectly rational for a subsistence farmer to clear a patch of rain forest so he can feed his family, the act is irrational from a global perspective because it diminishes the future for all of us. Benefit to the few and cost to the many is a fundamental principle of natural resource exploitation, one dubbed the "tragedy of the commons" by Garrett Hardin.[14]

As wildlands are progressively appropriated for human use, less and less remains of the natural habitat required to support biodiversity. Where wildlands are extensive, as in Amazonia, their value tends to be very low. Wilderness is typically undervalued by market forces.

But with increasing scarcity, the value of wildlands, as of most commodities, tends to rise, increasing the opportunity cost of creating additional parkland. It is a vicious cycle that has already foreclosed the prospects for conserving many species that live in overpopulated parts of the globe.

A few thousand years ago, all land was wildland. Now, little remains. Most of the temperate regions of the earth have been preempted for human use. Half of the original tropical forest is already gone. Existing wildlands will continue to be nibbled away, largely for three reasons. First, most remaining wildlands officially belong to national states but in practice are viewed as unused land, available to the first claimant. Governments often pay scant attention to what is transpiring in remote parts of the country. Second, wildlands are often the only uncommitted lands available for expansion of forestry, agriculture, mining, or livestock husbandry or simply for accommodation of a growing population. They are consequently under mounting pressure as economic expansion notches up the demand for commodities. Third, and most compelling, wildlands are reservoirs of unexpended resource capital, whether it be in vegetation, soil, or subsoil.

In other words, wildlands epitomize the last free lunch. Trees hundreds of years old growing on fertile soils that have formed over thousands of years are up for grabs by the first comer, who reaps instant rewards. No one has had to invest in the land or wait for the trees to mature. No replacement cost needs to be paid. And the gains to be made can be enormous. Eons of accumulated resource capital lie free for the taking. As a result, there can only be continued losses of wildlands wherever they still exist.

Once exploited, wildlands typically experience a period of extensive use. Extensive use, in contrast to intensive use, is economic activity that does not completely destroy the natural vegetation cover. In North America, trappers preceded loggers in the wilderness, and loggers were often followed by settlers. Trappers removed only animals, leaving the forest intact. Loggers left a depleted forest, but it was still a forest. Settlers then removed the remaining trees, thereby crossing the threshold between extensive and intensive use. A similar progression was followed on the Great Plains, where cattle ranchers preceded the sodbusters. In the Tropics, forests are commonly used for hunting, then selective logging, and then slash-and-burn agriculture. The latter is counted as extensive use because 90 percent or more of the land is fallow at any time and remains vegetated with native species, even though biodiversity may be severely diminished.

Tropical wildlands continue to be invaded by a steady stream of land seekers who incrementally push back the frontier. As the frontier recedes farther and farther from existing roads, settlers whose lands lie beyond the reach of transportation clamor for more roads penetrating even farther into the wilderness. There is no logical end to this process other than complete liquidation of the wilderness.

Once the frontier has passed over the horizon, the land faces one of two fates. If it is suitable for mechanized agriculture, cattle ranching, or tree plantations, it will be appropriated for one of these more intensive uses. If the land is too remote, steep, rocky, or barren to support intensification, it may remain in slash-and-burn cultivation. Potentially a stable form of land use, slash-and-burn agriculture becomes unsustainable when demographic pressure forces farmers to decrease rotation times in an effort to produce more from the same land. Declining yields precipitate a downward spiral, at the bottom of which lies human destitution and abandonment of the exhausted land.[15] The squalid cities of the developing world are replete with throngs of disillusioned farmers who were forced to either abandon their land or face starvation.

Land abandoned by slash-and-burn farmers may experience one of several fates. It may be left to recover as second-growth forest, in which case it remains in the category of extensive use. Or it may be transformed by fire into grassland, as is happening across huge areas of the humid and seasonal Tropics. There, local people customarily set fire to the grass every dry season to prevent the reestablishment of woody vegetation and to perpetuate the grassland. In Southeast Asia, fires have encouraged the growth of alang-alang *(Imperata cylindrica)*, a coarse, inedible, fire-tolerant grass that has created wasteland out of millions of hectares of land.[16]

Economics inevitably favors intensification of land use wherever the land can support it and, often, even where it cannot. Farming replaced cattle ranching in the tallgrass prairie of the United States because it brought a higher return. In the southeastern United States, tree plantations are more profitable than natural forests and are rapidly replacing the latter. I strongly suspect that the same will happen to much of the land now occupied by tropical forests.

The total amount of land under intensive use worldwide will continue to grow slowly into the twenty-first century but with relatively small gains in cropland and pasture.[17] Because nearly all suitable land is already in use, new additions to the global stock of cropland will fail to offset losses to urbanization, salinization, and other factors. For example, the total area of grainfields harvested worldwide is dwindling, having decreased by more than 5 percent since it peaked in 1981.[18] The additional grain needed to offset population growth has come entirely from productivity gains, but the margin for further improvement in productivity decreases with each passing year. Grazing lands are already so overcommitted worldwide that global production of beef and mutton began to drop in 1990, having increased throughout the century. Decreased production is attributable to systematic overgrazing and consequent land degradation. The largest gains in the intensive use category are to be realized through reforestation efforts and the conversion of natural forests to tree plantations. Much of the latter will register as decrements in the current areas of wildland and land under extensive use.

One consequence of the land-use cascade is certain: wasteland will continue to accumulate through the coming decades. Additions to wasteland will come from many sources: erosion of existing cropland and pastures, soil exhaustion, salinization, aquifer depletion, desertification, and, increasingly, urbanization.[19] Wasteland is accumulating far more rapidly than it is being restored to productive use, not because of a lack of scientific know-how but because there are few economic or political incentives for rehabilitating wasteland so long as undeveloped lands remain.

What are the keys to stabilizing land use? First and foremost, the human population must be stabilized. The term *sustainable development* will remain an oxymoron until population growth ceases and the land-use cascade can be halted or, better, put into reverse. Europe, for example, has achieved a stable population and is not experiencing rapid changes in land use. Countries still experiencing rapid population growth should implement national land-use plans. Such plans are needed to preserve wildlands and to prevent the steady accumulation of wasteland. Many countries, including some in the Tropics, have classified their national territories for appropriate uses but systematically ignore these classifications. Creating a plan is easy; the big hurdle is implementation.

Implementation can be achieved through zoning or through systems of incentives and disincentives.[20] Each has advantages and disadvantages with respect to the other. Zoning is more rigid, limiting flexibility of responses to changing economic conditions. Systems of incentives and disincentives are more flexible, but they are also more subject to political manipulation and can be overridden by unanticipated economic swings. Creation of either system is preferable to the laissez-faire approach to land use embodied in the frontier mentality.

Unfortunately, the frontier mentality, now a glaring anachronism, dies hard.

Looking forward, let us imagine a country that has set itself the task of designing a comprehensive land-use plan. Will there be a provision in the plan for wildlands? Wildlands, by definition, are lands that yield no direct economic benefit, except perhaps via ecotourism. Will leaders push for preserving such lands? If the country is already overpopulated, the answer is certainly going to be no. With the notable exception of Costa Rica, I cannot recall ever hearing a politician from a developing country talk about wildlands except in the presence of representatives of international donor organizations who were there to offer him money for preserving wildlands. For many societies, wildlands are an expedient, not a cultural asset.

Few politicians in the developing world have a vision for their country that includes wildlands. The existence of wild, uninhabited, roadless areas runs counter to the image of a developed country. In dreaming about the future, politicians in Brasília, Lagos, or Jakarta have in mind images of their visits to Europe or the U.S. heartland. The vision is one of expansive and productive farmlands linked to vibrant industrial cities by busy expressways. Mental images of Yellowstone National Park, the Grand Canyon, or Denali National Park aren't part of the picture. Worse, wildlands are viewed as a national embarrassment, a mark of underdevelopment and immaturity as a nation. The object is not to cherish and preserve wildlands but to convert them to the vision they associate with prosperity, be it the tidy farmlands of Europe or the vast urban-industrial panorama of a Chicago or Los Angeles.

Political attitudes toward tropical forests are revealed in statistics on forestry concessions, that is, logging rights granted to timber interests. In some countries—Thailand, the Philippines, Ghana—the area under concession exceeds the remaining area of productive for-

est.[21] The contradiction arises because lands still registered as concessions have illegally undergone deforestation. There is no expectation
of future wildlands here. In some other countries, namely, East
Malaysia, Indonesia, and Bolivia, forestry concessions already cover
more than three-quarters of remaining production forest.[22] In still
others—the Democratic Republic of Congo (formerly Zaire), Perú,
Colombia—forestry concessions have been granted in only a minor
fraction of the countries' forests, but only because the timber boom
has not yet arrived.

When the boom does arrive, it can unleash an all-out assault on a
region's timber resources. Not long ago, I was stunned to hear a pre-
sentation by conservation biologist Christopher Uhl of Pennsylvania
State University in which he described the frenzy of activity that had
engulfed the *várzea* (floodplain) forest at the mouth of the Amazon
River. He presented a map showing the locations of 1,300 active
sawmills in a relatively circumscribed area around the Brazilian city of
Belém. Eighty-five percent of the wood was being used to supply the
country's domestic market. Uhl predicted that the exploitable timber
of the *várzea* would soon be exhausted and that when it was, the logging boom would begin a long march up the Amazon, consuming
everything of value in its path.

One encouraging sign is that a number of tropical countries, Perú
and Indonesia among them, have classified their forestlands in a way
that distinguishes between production forests and protection forests.[23]
Production forests, as the name implies, are designated for exploitation, usually via concessions to private interests. Protection forests are
deemed too environmentally sensitive to withstand exploitation and
are excluded from concessions. Of course, many of these are forests
of little commercial value growing on barren or rocky ground or,
more commonly, are forests growing on steep slopes. In this case, provision of an environmental service—watershed protection—can jus-

tify the maintenance of wildland, but only where the commercial potential is minimal.

Protection forests are a step in the right direction, but what will happen in the future when pressures on natural resources increase still further? Will concession holders honor the distinction between production and protection forests then? Only if there are competent and dedicated enforcement agents to hold the line. Even then, holding the line will depend, more than anything, on whether the general level of prosperity rises or falls. If it rises, there will be hope because in a vigorous economy, the incentive to take bribes diminishes. Where prosperity falls, there will, I fear, be little hope. The pressure on natural resources will simply be too great to resist.

Stabilization of land use depends not only on whether there is a land-use plan but also on who owns the land. In the United States and in the West in general, there is a tradition of private property rights that imposes few or no constraints on the current owner. In the era of the frontier, when land was available for the asking, the notion of absolute property rights was understandably popular. Anyone who acquired rights to a tract of land could exploit its resources without restriction. Land acquisition thus offered a route to wealth. What politician in the land of opportunity could espouse closing off a route to wealth?

The U.S. tradition of absolute property rights will eventually be recognized for what it has already become, a detrimental anachronism, an impediment in the path of real progress toward sustainable development. The point can be made in the form of a simple ethical conundrum. The land belongs to the planet earth and is here forever. An individual lives for but a fleeting moment in time. Is it then right that an individual be allowed to degrade the land, making it unusable by future generations? A doctrinaire capitalist would say yes. If the land is degraded, its price will drop and the owner will be obliged to absorb the loss on sale of the land. Fair enough. The logic of economics is unassailable, but it is amoral.

The response cannot be generalized. If every landowner on earth were to destroy his or her land, we would be the inheritors of Babylon. What economics fails to consider is that we are responsible for the earth that nourishes us and for the generations to follow. Yet we allow economics to run our lives and to dictate our laws.

An economic system that espouses unending growth, discounts the future, and undervalues natural resources is diametrically incompatible with sustainable development. One cannot rationally believe in them both. Yet under a morality in tune with a truly sustainable society, the idea that an individual could buy, own, and destroy for profit a forest of thousand-year-old trees would be completely abhorrent and unacceptable.

Who, then, should own the earth's remaining wildlands? At present, wildlands are owned primarily by governments, but often with a lien in the form of a timber concession. My cautious opinion is that the future of wildlands will be brighter if they remain under government control. To make the point, I offer two examples from the United States.

The timberlands of the Pacific Northwest (Oregon and Washington) are owned in part by private interests (35 percent), in part by state and local governments (7 percent), and in part by the federal government (57 percent).[24] Private owners have systematically harvested the old growth on their lands and have replaced it with even-aged monocultures supporting a sharply reduced diversity of other taxa. Federal lands, although now largely ravaged by clear-cutting, in principle fall under tighter restrictions, including those specified in the Multiple Use–Sustained Yield Act of 1960, the Endangered Species Act of 1973, and the National Forest Management Act of 1976.[25] One stipulation is that the harvest in a given year not exceed net growth, yet in 1986 the harvest from national forests in western Oregon and Washington exceeded net growth by 70 percent.[26] Clearly, we do not always adhere to our own laws, especially when managers are under political pressure.

Nevertheless, more than 90 percent of the remaining old-growth forest in the Pacific Northwest is on federal lands (amounting to less than 10 percent of the original forest), and the battle over old growth is entirely about the policies to be followed by federal managers. Private forests hardly enter the picture. It is safe to say that had there not been federally owned forests in the Pacific Northwest, there would be no old growth today outside formal parks. Even so, the battle rages on, and the end is not yet in sight.

My second example is from Texas. Texas is unique among the states in having briefly been an independent republic before it joined the nation. As a republic, Texas negotiated the terms for becoming a state. One of the conditions won by negotiators for the Republic of Texas was that all land in the state would be privately owned. Today, as a result of these negotiations, only a relatively minuscule 3 percent of Texas is publicly owned land.[27] The effort to conserve the biodiversity of Texas has consequently been extraordinarily difficult. Highly endangered endemic species, such as the black-capped vireo and the golden-cheeked warbler, occur almost entirely on private land, engendering uneasiness or hostility among landowners toward the Endangered Species Act.[28] The option of ordering federal land managers to alter their practices to favor endangered species is not available on private lands. If endangered species are to be protected, either landowners must accept restrictions on the use of their property or private conservation groups such as The Nature Conservancy must buy large tracts of land. Neither of these alternatives is attractive when the burden falls disproportionately on local residents.[29]

If one were to ask the innocent-sounding, but in reality very difficult, question "Where does one find the biodiversity of the United States?" the short answer would surely be "On federal lands." Federal landholdings tend to be large, and they tend to be managed less intensively than private lands. Increasingly, federal land managers are being required to manage for multiple use and to cater to the growing public demand for recreational opportunities and protection of biodiversity.

The reason so much biodiversity is concentrated on federal lands in the United States is that federal law prohibits changes in land use. A national forest is a forest by law. A national grassland is a grassland by law. Never have most of these lands been anything but what they are now. Abused? Yes. Degraded? Often. But most federal land has never been plowed or planted; hence, it retains disproportionate value for biodiversity. In contrast, private land is extremely fragmented and subject to changing land use.

The wonderful thing about the U.S. system of land tenure, if I may indulge in some unabashed chauvinism, is that it has conserved most of the country's biodiversity. This fortunate eventuality has come about largely by accident, not by intent. Regardless, the history of land use in the United States convinces me that the nation stumbled fortuitously onto a highly desirable mix of public and private landownership. Most private land is intensively used for the maximization of profit; public land, by law, remains under extensive use. It is the public-private split in land use that has prevented the United States from stumbling down the land-use cascade. Thus, the United States is that much closer to achieving one criterion of sustainable development—a stable pattern of land use—than are many other countries.

Few people seem aware that the United States is nearly unique in its mixed system of public and private landownership. I have talked about this issue with many people, including professional conservationists, and have been surprised to find that few of them have ever given it a moment's thought. U.S. citizens take their system for granted and usually do not look beyond their own shores.

In fact, however, few countries have formally set aside much public land, and even fewer have enforced the permanency of land use. Tropical countries that continue to retain forests in the state sector are mainly hedging their bets because state control of forests allows politicians to manipulate an enormous source of wealth, the timber resources of the primary forest. By wielding control over forests and other natural resources, government officials are in a position to accumulate vast wealth through illegal means. In countries where political office opens opportunities to share in the spoils of pillage, there is no commitment to management, only to exploitation.

Yet a nation's prosperity or poverty ultimately hinges on the way its natural resources are conserved and managed. Ideally, natural resources should be shared equitably among a nation's people and conserved for future generations. In practice, and I do not exempt the United States, resources are too often a magnet for speculators and exploiters.

Serious resource management requires serious and competent government. Few tropical countries are so blessed. Still, given a choice between public and private ownership, I would favor the former, if for no other reason than that governments are often lethargic and incompetent and thus unable to wreak much damage. Turn a country's natural resources over to private interests and the consequences are predictable.

Although human beings have been waging wars of conquest since before the dawn of history, the United States has been slow to recognize the relevance of natural resource policies to international relations. On April 12, 1996, U.S. Secretary of State Warren Christopher gave a major policy address at Stanford University in which he announced a new definition of national security, signaling a significant shift in the objectives of U.S. diplomacy. From then on, he proclaimed, the environment would become a top priority at the Department of State, assuming an importance on a par with that of preserving peace.[30] Indeed, the environment and peace have become inseparable issues in many parts of the globe.

Christopher's epiphany is both a welcome nod to reality by the U.S. government and an opportunity to redirect foreign assistance programs toward more productive ends. If Secretary Christopher and his successors are truly committed to sustainable development, they would do well to encourage governments to develop national land-use plans and to enact measures designed to stabilize land use in accordance with rational, scientific criteria. The conspicuous failure to date of foreign assistance programs in developing countries has resulted in widespread "donor fatigue."[31] New approaches based on the maxim that nothing is more precious than the land and its resources might begin to make a difference.

In the United States and in much of the rest of the world, it is clearly individuals, often acting collectively in the form of corporations, who profit most from the exploitation of natural resources. It is just as clear that individuals in unencumbered pursuit of the profit motive bring impoverishment to whole societies when resources are exhausted. Intergenerational equity is not a popular theme among economists, but it should be a matter of the greatest interest to parents concerned about the welfare of their children and grandchildren. After all, economists have children, too, and thus they have a compelling incentive to put aside their discount rates and think more rationally about the future.[32]

10

WHY CONSERVATION IN THE TROPICS IS FAILING:

The Need for a New Paradigm

TROPICAL PARKS ARE failing in country after country because the institutions created to protect them are weak and ineffectual. Institutional weakness derives from many sources, but it arises ultimately from the low priority governments give to protection of parks. So much has been recognized since tropical parks began drawing the attention of the international conservation community. Conservation organizations have responded by providing assistance, but over time, as a result of frustration born of failure, the form of assistance has changed dramatically, and not necessarily for the better.

In the early 1970s, international conservation assistance took the form of donations of tangible assets such as vehicles, boats, motors, guard posts, uniforms, generators, and radios. Such assistance brought visible benefits, without which parks often could not function. By helping to put parks on an operational footing, the proffering of material goods directly addressed the challenges of protecting biodiversity. Manu National Park, for one, would have remained a paper park without direct material assistance.

But in the face of mounting pressures, tangible donations proved less and less effective. Lying behind the donations of equipment was the unstated assumption that improved transportation and communications would promote effective enforcement of park regulations. The model in the minds of donors was that of the National Park Service in the United States and its highly effective rangers. Unfortunately, all too often the model proved to be out of synch with the realities of the recipient country, and efforts to replicate it ended up being little more than an expression of the donor's cultural bias.

Having well-trained and well-equipped guards is an indispensable prerequisite to creating an operational park, but the presence of guards amounts to more show than substance if they cannot make arrests and if the police and courts do not provide the necessary backup. Guards that are not empowered, by default, occupy themselves with controlling tourists and scientists while closing their eyes to the more serious challenge of evicting squatters, loggers, and miners.

Conservation organizations have yet to devise a workable strategy for combating illegal activities in tropical parks. Law enforcement is not an area in which the participation of foreign organizations or their local surrogates is welcome. Law enforcement, to the extent it is carried out at all, is the cloistered province of police and military forces that are highly jealous of their authority and autonomy. In many tropical countries, civilian politicians have little or no ability to influence the uniformed services. A national parks agency with the best of intentions remains powerless without the backing of those who carry the guns.

Unable to influence the layers of society that hold all the cards, international conservation workers find themselves facing a gargantuan task in their efforts to oppose the forces that threaten parks. Awareness of this reality has gradually sunk into the consciousness of conservation professionals, who remain frustrated with their inability to exercise control over events on the ground. How can the integrity of parks be ensured when effective law enforcement barely exists in many countries?

In some circumstances, the only recourse may be to solicit intervention at the highest level. An anecdote will illustrate the point.

A colleague of mine received an urgent communication from an associate in Indonesia notifying him that a logging company had begun to harvest trees in his study area inside Sumatra's Gunung Leuser National Park. The matter was complicated by ties between the regional military commander and the logging company. Knowing this, the park's administrator was powerless to do anything.

By chance, my colleague happened to be well connected in Europe. He consulted with some acquaintances in the World Wide Fund for Nature, and it was decided that he should draft a letter to President Suharto of Indonesia, drawing his attention to the violation of the park and the damage that logging would do to an international scientific enterprise of long standing. The letter was transmitted to a prominent European prince, who sent it to Suharto through diplomatic channels under his own name. Shortly afterward, the logging abruptly ceased and the company withdrew its equipment from the park. However, although this strategy proved effective in this particular instance, the use of international leverage to call for presidential intervention in a park enforcement crisis is a measure of last resort. It might prove effective the first time, but it would fail entirely as a general mechanism for combating violations.

Frustrated by its inability to grasp the stick of enforcement, the international conservation movement turned in the 1980s to the carrot of economic assistance. It was reasoned that local villagers might voluntarily respect park boundaries if they could raise their standards of living through means other than exploitation of a park's resources.

During the same period, the major international donors, led by the World Bank, were under pressure to adopt "greener" policies and lending practices. U.S. President George Bush promoted the greening of USAID, hoping to give credibility to his claim of being an "environmental president," even as he opposed most environmental reform at home. USAID's European counterparts were right in step.

The wispy notion of sustainable development provided the common ground for a joining of forces by the international banks and bilateral assistance agencies and the world's leading conservation organizations. The big donors knew little about sustainable development, but they wanted to look green. Conservation organizations, impeccably green but inexperienced in managing international development projects, wanted to expand their programs and their international reach. Through multimillion-dollar USAID projects, conservation groups could demonstrate impressive rates of financial growth to their boards of directors.

The self-interest of the two sets of organizations converged in the form of integrated conservation and (sustainable) development projects, called ICDPs. The stated purpose of many ICDPs is to reduce external threats to parks by promoting sustainable development in surrounding areas. Conservation organizations were enthusiastic about ICDPs because the title contains the word *conservation,* and bilateral aid agencies liked them because the title contains the word *develop-ment*. Each side could frame projects in its own image.

Frequently, but not always, the focal points for ICDPs were pro-vided by biosphere reserves established under the auspices of the Man and the Biosphere Programme of the United Nations Educational, Scientific, and Cultural Organization (UNESCO). To qualify for UN-ESCO recognition as a biosphere reserve, a park must have regional significance as a repository of biodiversity, must possess an inviolate core, and must be surrounded by one or more multiple-use areas serving as buffer zones for the core. Although ICDPs promote sus-tainable forms of development in the buffer zones, the projects fre-quently do not directly involve the park itself, on the ground that the park is the government's responsibility.

For USAID and its counterparts, ICDPs provided a green shield under which they could continue what they had been doing for dec-ades, sponsoring rural development programs in developing coun-tries. For the conservation organizations, ICDPs offered a path to rapid growth. In 1995, one leading U.S. conservation organization al-located more than 60 percent of its program budget to various activi-ties under the rubric of "Conservation and Human Needs"—ICDPs by another name. Much of the funding came from USAID or other major donors.

Despite the gilded rhetoric presenting them as conservation en-deavors, ICDPs represent little more than wishful thinking. Project objectives typically have little direct relevance to the protection of biodiversity.[1] To the contrary, project managers who successfully in-novate and invigorate the local economy risk aggravating the very problem they are trying to solve. By stimulating the local economy, an ICDP attracts newcomers to a park's perimeter, thereby increasing the external pressure on the park's resources.

More fundamentally, the ICDP model itself is inappropriate. Whereas in principle, parks are intended to be permanent institutions, projects, by their very nature, are designed to be executed within a fixed term, usually three to five years. Money floods in at the beginning and then stops just as abruptly at the end. Activities supported by the money stop when the money stops. All too often, the conditions the project was intended to ameliorate revert to their former state when the project ends, with little or nothing accomplished.[2]

There are many other drawbacks to ICDPs as an approach to the conservation of biodiversity. Most basically, ICDPs shoot at the wrong target—local people instead of the park and its natural resources.

Rural development projects characteristically emphasize intensification of land use through agroforestry schemes, new crop varieties, enhanced methods of animal husbandry, improved land management practices, small-scale irrigation works, preliminary processing of products, and so forth. Rarely is it acknowledged that intensified methods and improvements in rural standards of living also stimulate increases in population density. A farmer whose income rises may well hire additional labor. Increased cash flow through a community will attract trickle-down businesses such as stores, transportation services, video parlors. The outcome? More people are drawn to the community at a time when there are already too many people living around park boundaries. That is the problem. Simply put, successful ICDPs are only likely to make population-related problems worse.

Another crucial problem with ICDPs, one that is rarely addressed, is land rights. Forestlands often belong to the state and are used on an informal basis by indigenous slash-and-burn agriculturists and recently arrived settlers alike. How different was the settlement of North America! Following cultural practices inherited from their English forebears, government agents sent surveyors into the wilderness to measure and map the land prior to its settlement. Settlers then bought or registered their land before investing in buildings, clearing, or making other so-called improvements.

Such a degree of forethought and organization is unimaginable in many tropical countries. Forestlands are rarely held under title, or, if so, the titles are held by wealthy absentee landlords or speculators, as in Brazil. The bureaucratic process of titling a parcel can be so arcane and labyrinthine as to be incomprehensible to a semiliterate peasant. Thus, in the absence of any legal alternative, an attitude of lawlessness toward the land has evolved. Lawlessness is honored in the breach by politicians unwilling to risk social unrest by depriving poor people of access to land.

Unlawful occupancy of land is simply a fact of life throughout Latin America, in both rural and urban settings. For example, much of the growth of the sprawling barrios and *favelas* of Latin megacities such as Mexico City, Caracas, Lima, and São Paulo has occurred through *invasiones*. Landless migrants to the city gather together, target a vacant plot, and, on an appointed night, rush in to erect makeshift shanties of cardboard and sheet metal before dawn. The next morning, the landowner is faced with a fait accompli. Appeals to the police are of no effect because the politicians have instructed them not to interfere. Politicians who are regarded as lenient toward land invaders can help themselves to the votes of the poor, and thus they view squatters as an easy way to expand their constituencies. And so the cycle continues.

In societies in which extralegal land occupancy is the norm among the poor, one cannot simply draw an invisible line around a tract of forest, declare it a park, and expect the line to be respected. By custom, unoccupied land is available for the taking. Such a long-standing practice cannot be overturned over the course of a three- or five-year rural development project. Moreover, land titling is a function reserved for the state and is thus beyond the practical scope of an international assistance project. These facts of life lie behind the failure of many ICDPs to strengthen park boundaries.

Yet another shortcoming of ICDPs is that they typically reflect a community's existing economy and infrastructure. But many developing countries are undergoing rapid change. Common project goals, such as establishment of agroforestry plantings, installation of micro-irrigation works, and other such facets of rural development, can be rendered irrelevant by suddenly changing economic conditions.

The most profound change to a rural community comes when the government builds a road connecting it to a major city. An entire spectrum of new development possibilities then arises: large-scale logging, intensification of agriculture, production of new cash crops, cattle ranching, and so forth. If the area has agricultural potential, land values can soar, and many of the original inhabitants will be bought out by expanding agribusiness. In short, the tacit assumptions that underlie a project's design and goals can go out the window overnight.

Parks are also affected by forces that lie entirely beyond the reach of ICDPs. Whereas village poachers and slash-and-burn agriculturists mainly nibble around the edges of park boundaries, doing minor damage, large logging and mining companies can inflict damage on a massive scale. Invasion of parks by big-time resource pirates is nearly always sanctioned (for a cut in the proceeds) by the local military commander or by influential power brokers in the capital city. People such as these are untouchable at the village level. They can be stopped only through appeals to the head of state, if they can be stopped at all.

For all these reasons, ICDPs, in my view, are an inappropriate response to the external forces that threaten parks. I see these projects as misguided efforts that mostly fail to advance conservation, regardless of any success that may be achieved on the development side. Moreover, I doubt that it will ever be possible to improve ICDPs to make them more effective as instruments of conservation. The basic concept is flawed. Meanwhile, millions of dollars are being poured into ICDPs in the name of conservation.

Let me be clear. I have no objection to ICDPs per se. My objection is to the illusion created by linking them to conservation. Instead of trying to promote economic growth around parks, it would be better to discourage people from settling in or near buffer zones, perhaps by persuading governments not to build roads in these areas. If there are to be ICDPs, they should be located at a distance from parks so that people might drawn away from park perimeters rather than attracted to them.

Another misconception embodied in the ICDP concept is the conviction that social change can be brought about through bottom-up processes. The ICDP approach assumes that the destiny of parks lies in the hands of local people, an assumption that is only partly correct. What ICDPs do not take into account is that local people are only minor players in a much larger theater. The lives of village people are strongly influenced by decisions of the central government and conditions determined by it: construction of roads; availability of rural credit, subsidies, and tax incentives; inflation versus stability of the national currency; raising or lowering of trade barriers; laws governing labor practices; receptivity to foreign capital; and so forth. Against powerful forces such as these, ICDPs pale into utter insignificance.

Now we have come full circle. Unable to grasp the stick of enforcement, conservation organizations turned to the carrot of economic assistance, but they must now come to grips with the failure of that approach as well. Bottom-up processes initiated at the village level will not improve the security of parks because they rely 100 percent on voluntary compliance. The recent history of environmental legislation in the United States provides many examples of the failure of voluntary compliance to produce the desired results. There are cheaters in every society, and the only response to cheating is enforcement. There is no substitute for enforcement. Without it, all is lost.

The focus of conservation must therefore return to the make-it-or-break-it issue of actively protecting parks, a matter that hinges above all on the quality of enforcement. Active protection of parks requires a top-down approach because enforcement is invariably in the hands of police and other armed forces that respond only to orders from their commanders. When the commanders happen to be business partners of the local timber baron, the prospects for protecting nature are undeniably grim.

11

HARD CHOICES
IN THE TWENTY-
FIRST CENTURY

THE TWENTY-FIRST century will be a time
of unprecedented change, accompanied by unprecedented social tur-
moil. Given the numerous imponderables that cloud the future, it is
impossible to see far ahead. People living today, myself included, simply cannot imagine an earth occupied by 10 or 12 or 14 billion people.
The 5.8 billion alive today are already far too many.

The world's economy currently demands more fish than the seas
and freshwater together can produce. The earth's grasslands have
peaked in their capacity to supply beef and mutton. The demand for
wood, paper, and other forest products is rising exponentially and
shows no sign of leveling off.[1] Nearly all arable land is already in use.[2]
Dams and diversions for irrigation have altered 77 percent of the total
water discharge of the 139 largest river systems in Europe, northern
Asia, and North America, imperiling scores of species of fish, mol-
lusks, and crustaceans.[3] Underground aquifers are being overpumped
in nearly all the earth's arid zones.[4]

An additional doubling of the earth's population will result in un-imaginable stresses on the world's renewable resources. Although no one can predict exactly what to expect on the social front, one can hardly doubt that the world will become an even more volatile and dangerous place than it is today.[5]

With increasing pressure on the world's renewable resources, we can anticipate a pattern of exploitation that follows a hierarchy of sub-stitutions. The processed fish sold in small, squat cans exemplifies my point. When I began research in Perú in the 1960s, the cans were la-beled *atún* and contained bluefin tuna. But over the years, the name on the cans has changed: from bluefin tuna to yellowfin tuna to bonito, to albacore, to mackerel, to jurel, and, finally, to sardines. Each name change signaled the exhaustion of a resource. The lowli-est species is reserved for last, but in the end, it is used. Such substitu-tions mark a steady erosion of our quality of life, if not a decrease in the total volume of resources entering the world economy.

The same principle of substitution applies to forest resources, whether timber or game. At the edge of the Amazonian frontier, hunters disdain smaller prey in favor of deer, tapirs, peccaries, and the larger monkeys, especially woolly and spider monkeys. Behind the frontier, near established settlements, these premium species have long since been extirpated and hunters concentrate on lesser fare, such as rodents and armadillos. The high diversity of tree species in tropical forests invites an almost infinite substitution: first mahogany, then cedar, then certain legumes, and on and on. When demand is in-satiable, it will drive the exhaustion of every usable species, no matter how humble its properties.

Short of miraculous transformations in the attitudes of govern-ments, the earth's remaining old-growth forests are destined to disap-pear, and along with them, their biodiversity. The amount of old-growth forest remaining in Europe and the eastern United States represents roughly 1 percent of the original forested area. Canada still retains half its primary forest, but nearly all of that is in the unpopulated north.[6]

The vast scale on which surviving old-growth forest is being liquidated is not widely appreciated. With the dissolution of the Soviet Union, the vast virgin forest of the Russian far east has been opened for exploitation by a government desperate for foreign exchange. Hyundai Group, the Korean conglomerate, clear-cut 150,000 hectares in the region of Amur Oblast (Province) in 1993 alone, significantly reducing the habitat of the world's 200 remaining Siberian tigers.[7] Even the U.S. Department of Commerce has joined the rush to get in on the Siberian timber bonanza, granting $500,000 to the Global Forestry Management Group, a consortium of Pacific Northwest sawmills, to study the feasibility of logging millions of hectares of old-growth forest in Khabarovsk Kray, a territory of eastern Russia.[8]

Within the United States, the picture is little better. Years of effort by relatively influential conservation groups have failed to halt logging of the last remaining old-growth forest of the Pacific Northwest. Strong public opposition to further clear-cutting is revealed in opinion surveys, yet congressional delegations from the states of Washington and Oregon unanimously supported the timber industry throughout the 1980s. The reason for this apparent anomaly is obvious: industry-supported political action committees contributed six times more to congressional campaigns during that period than did conservation organizations.[9] If the wealthy and democratic United States cannot halt the ravaging of its last old-growth forests, how can poor developing countries be expected to do so?

Against the collusion of political and industrial powers, public opinion has little relevance, even in the most democratic societies. Consider the case of British Columbia's Clayoquot Sound, until recently one of the last untouched watersheds in the Pacific rain forest. Canadian environmental groups tried for years to prevent clear-cut logging of its forests, yet:

> In April 1993, the British Columbia government announced its decision to allow logging in 74 percent of the Clayquot's ancient rain forest, a 650,000-acre mosaic of towering cedar and thousand-year-old Sitka spruce on the western shore of Vancouver Island. Only weeks prior to this decision, Premier Mike Harcourt's provincial government had invested $50 million into MacMillan Bloedel, Canada's most notorious industrial logging juggernaut, becoming the company's largest shareholder. . . .
>
> That summer, having exhausted all other avenues of protest, environmentalists pursued the honorable tradition of non-violent civil disobedience. By summer's end, ten thousand people had passed through the protester's camp . . . [registering] the largest protest in Canadian history.[10]

Unlike many conservation stories, this one has a relatively happy ending. After more than 1,000 protesters had been sentenced to jail terms, the provincial government took notice of the political risks it was running in promoting the clear-cutting of Clayoquot Sound's ancient forests. As I write this, in late 1997, the rate of cutting has been reduced from 600,000 to 40,000 cubic meters per year. MacMillan Bloedel Ltd. is out of the picture, and indigenous communities are managing a far less damaging harvest through single tree selection.

In countries with less developed democratic traditions, the power of ordinary people to influence government decision making can be minimal. In 1991, the small South American country of Guyana granted a fifty-year license to the Barama Company Ltd., a Malaysian corporation, to exploit 1.69 million hectares of forest occupied by

1,200 Amerindians, who hold no formal title to their lands. Barama obtained extremely favorable terms, amounting to a giveaway of the country's resources. It was granted a ten-year tax holiday during which it will not pay income tax, corporate tax, withholding tax, consumption tax, property tax, or export or import duties.[11] Although the concession amounted to 9 percent of the national territory, public participation in the decision to grant it was nil.

Another example reveals the destructive vestiges of colonialism. In a sardonic twist on the debt-for-nature theme, the French government agreed to cancel half the debt owed by the government of Cameroon in exchange for a deal giving French companies nearly exclusive rights to log the country's forests. A major beneficiary is likely to be the Société Forestière Industrielle de la Doumé, of which Jean-Christophe Mitterrand, son of the former French president, is a leading figure. The French offered to cancel even more debt on passage of a bill by the Cameroonian parliament authorizing an increase in the area of forest open to logging by nonstate companies. A Tropical Forestry Action Plan, drawn up in consultation with United Nations agencies, set targets that would make Cameroon Africa's largest timber exporter by the year 2000.[12]

The governments of the United States, Canada, and France, as well as the United Nations, pay lip service to the notion of sustainable development, but their actions belie any serious commitment. Decisions are made ad hoc, on a case-by-case basis, and often under political pressure. What is lacking is any plan, any vision of the future. In the absence of explicit long-range goals for a given country or for the world as a whole, there is nothing to restrain incremental destruction of the environment. The exigencies of the moment will always override a vague concept such as that of sustainable development unless the concept is given teeth in the form of explicit and enforceable objectives.

There is no time to lose in setting those objectives. Tropical landscapes are being transformed at a startling rate. Consider the following examples.

In 1970, botanist Calaway H. Dodson of the Marie Selby Botanical Gardens in Sarasota, Florida, founded a tiny biological research station at Río Palenque, in the lowland forest of western Ecuador. In the 1960s, much of that country's Pacific lowlands were still an unbroken carpet of virgin forest. The station was so isolated in its early days that reaching it from the nearest road required a three-day journey by horseback and canoe. But by the mid-1970s, the one square kilometer protected by the privately owned research station stood as the only remaining example of its forest type in existence.[13] Today, the Río Palenque Science Center contains the only known living specimens of more than thirty plant species. Explosive development of the region accompanying the construction of an all-weather road connecting the capital, Quito, to the main port, Guayaquil, made an island of the reserve in less than a decade. When I visited in 1985, the forest at Río Palenque struck me as a forlorn aberration in a sea of oil palm plantations and cattle pastures that stretched to the horizon.

A colleague of mine, Mercedes Foster, relates a parallel story about eastern Paraguay, where she conducted research between 1978 and 1983. At that time, there were only two trunk roads in eastern Paraguay, forming two sides of a triangle with the apex at the capital, Asunción. One road ran to Encarnación; the other, to Ciudad del Este. Lacking road access, most of eastern Paraguay was still cloaked in primary forest. The government, with financial support from the World Bank, began in 1979 to build a road to close the triangle. Traveling the completed road in 1995, I would never have known that the entire 320-kilometer route had been forested only sixteen years earlier. As far the eye could see, large-scale agribusiness monopolized the landscape in a manner reminiscent of the Midwest region of the

United States. The distinctive forest type of eastern Paraguay is now highly endangered.

With bulldozers, chain saws, and huge reserves of capital, humans can now alter the face of the earth on scales of space and time that were inconceivable even a generation ago. Remoteness has preserved wilderness in such places as Siberia and the Amazon, but soon remoteness itself will be a thing of the past. Renewable resources of all kinds will be exploited to exhaustion unless restrictions are vigorously imposed. But will such restrictions be imposed in time? To date, most governments have taken the contrary path, offering subsidies and other incentives to encourage resource exploitation in their short-sighted rush to economic development.

Whereas sustainable use of renewable resources undeniably contributes to economic development, resource exhaustion does not. This is an elementary distinction but one that has escaped the attention of many world leaders. Just as surely as restrained resource use brings prosperity, resource exhaustion brings poverty and social disintegration. Although the logic of this statement is simple and obvious, rarely does one see it spelled out in the media, much less in the context of continued population growth.

Although the media remain steadfastly in denial, there is at long last a dawning recognition among intellectuals, and even a few politicians, that overpopulation leads directly to resource depletion and environmental degradation and that loss of environmental capacity leads inexorably to social strife.[14]

Evidence in support of this simple chain of cause and effect abounds on the global stage, yet in reporting events, the press is extremely reluctant to delve into causes. As an example, I can recall my outrage on reading a six-page cover article in *Time* magazine on the second Ethiopian famine, in 1987.[15] The article described the grotesque suffering of the starving people, the mass movement of desperate refugees across international borders, the grim conditions of United Nations feeding camps, the sums of money being spent by international assistance organizations, the lack of government cooperation, and other poignant but tangential topics.

The question of causes was dismissed in a few glib sentences attributing the famine to drought. Nowhere did the article mention that Ethiopia is one of the most overpopulated countries on earth; that more than 90 percent of its once extensive forests have been cleared; that disastrous erosion has reduced the productivity of its soil; that even in a good year, agricultural production is barely sufficient for subsistence. Even if one were to deny that Ethiopia was overpopulated in 1987, at what point would it be considered overpopulated? The country's population, which was 21 million in 1950 and 51 million in 1990, is blithely predicted by the United Nations to reach 151 million by 2030.[16] Is such a number even conceivable? The thought stretches my imagination.

More recently, in the spring of 1994, I read thousands of words about the Rwandan genocide in such eminent outlets as the *New York Times* and the *Washington Post*. Occasionally, fleeting mention was given to the fact that Rwanda was "the most densely populated country in Africa," but without further elaboration. Not once did I read that Rwanda, with a density of 275 people per square kilometer and a population expected to double in twenty-five years, is full.[17] All arable land is claimed and being worked; the next generation has no prospect for acquiring land with which to feed itself. No wonder social tensions reached the point of boiling over. Garrett Hardin hit the target when he observed that in media accounts, "no one [has] ever died of overpopulation."[18]

The media carry a heavy responsibility in forming and arbiting public opinion. If the media fail to probe the most critical issues facing the world today, can the public be blamed for being unconcerned and for electing representatives who are outspokenly hostile to sustainable development? The media abrogate their responsibility when they deliberately avoid topics that might be construed as politically incorrect or offensive to some readers. Unfortunately for the world as a whole, overpopulation sits near the top of the list of taboo subjects. Instead of being suppressed, population issues should be on the front page every day.

Notwithstanding the end of the cold war, societal tensions continue to provoke conflicts around the world. To be sure, some are flare-ups of long-standing animosities that were squelched by authoritarian regimes during the cold war (as in Serbia, Croatia, and Bosnia) or are lingering by-products of the cold war (Afghanistan, Cambodia). Many others, however, pit one ethnic group against another independently of cold war influences, as with the Karens in Myanmar (Burma); the Muslims in Mindanao, Philippines; the Tamils versus the Sengalese in Sri Lanka; the Muslims versus the Hindus in Kashmir; the Kurds versus the Turks and Iraqis; the Hutus versus the Tutsis in Rwanda and Burundi; various tribes clashing in Liberia and Sierra Leone; and the Amerindians versus an oligarchic elite in Guatemala.

Estimates of the number of ongoing conflicts in the world range from 48 to 105, depending on the scale and intensity of hostility required to qualify for a given listing.[19] Regardless of how many are considered noteworthy, the fact remains that conflicts are active in a significant fraction of the world's countries. As fast as conflicts are resolved by mediation or intervention, new ones appear. It thus seems statistically inevitable that half or more of the world's nations will become embroiled in conflict over the next few decades. And that grim prediction ignores the likely rise in tensions associated with increasing environmental degradation and exhaustion.

War, whether civil or international, is not good for biodiversity. Any event that disrupts civil order is likely to unleash a run on unguarded natural resources, as has already occurred in a number of countries. In Chile, the shift from democracy to an authoritarian regime brought about unprecedented environmental change. Chile had been a democratic country for more than fifty years when General Augusto Pinochet overthrew President Salvador Allende, with connivance from the U.S. Central Intelligence Agency. Chile's temperate rain forest was then largely intact. Under Pinochet, generous logging concessions were granted to supporters and relatives. After democracy was restored, Chilean conservation organizations tried to stop the rapacious logging then in full swing, but they were blocked by court rulings declaring the concessions legal under Chilean law. Apart from the ironic question of how the decrees of an illegal government can be considered legal, the point is that without democracy, citizens have no means of holding a government accountable for environmental policy and actions.

At about the same time in Panama, actions of the U.S. government inadvertently began a wave of environmental destruction. Prior to the U.S. intervention, deforestation was causing so many negative environmental effects that the government of Manuel Noriega banned further clearing of forests. As soon as U.S. troops invaded to oust Noriega, throwing the government into disarray, peasants rushed to clear new slash-and-burn plots, even in the sensitive watershed of the Panama Canal. More than 50,000 hectares were cleared within a few weeks before order could be restored.[20]

In many regions of the world, social conflicts are financed by the sale of natural resources. Rebel armies in Angola and Mozambique were supported in part with receipts from the sale of ivory.[21] Rebels in the northern part of Myanmar finance their struggle against the military government with timber sold across the border in Thailand, as did the Khmer Rouge in Cambodia.[22] In the spring of 1995, Tutsi herdsmen returning to Rwanda from exile in neighboring Uganda drove thousands of cattle into Akagera National Park, usurping the only unoccupied space available.[23]

Cases such as these, and many more, warn that civil order and political stability are essential prerequisites for successful conservation. If, as I have surmised, civil strife is a future inevitability in at least half of the world's countries, the institutions charged with protecting biodiversity will be severely tested.

When new conflicts break out in tropical countries, will the international community intervene to restore order? The question is a current topic of debate in foreign policy circles. Recent history is mixed, indicating no consistent policy. The United Nations intervened to pacify conflicts in Cambodia and acted briefly in Somalia but not in Sudan, Rwanda, or Liberia. Fearing an invasion of refugees by boat if it did nothing, the United States invaded Haiti to oust a dictatorial regime and install an elected president. With extreme reluctance and after long delay, the North Atlantic Treaty Organization sent troops to enforce a negotiated truce in Bosnia. None of these incidents involved a major, strategically important country. To date, there has been no test case of a large, populous country gone berserk, although several are tinderboxes, among them Nigeria, Egypt, and Pakistan. It would be foolhardy to take bets. The United States, badly burned in Somalia and still remembering Lebanon, will be reluctant in the future to engage in altruistic adventures that entail significant political risk.

Gradually, the world will split into two camps: the countries that stabilize the size of their population and their use of natural resources and those that fail to do so. My hunch is that the world will increasingly avert its gaze as countries such as Haiti, Liberia, Rwanda, and Somalia fall into the black hole of chaos, as brilliantly foreseen by Robert Kaplan in his riveting essay "The Coming Anarchy."[24] A nation under the control of marauding bands of disaffected young men carrying automatic weapons will not conserve its natural resources, parks, or anything else. So far, we have seen only the tip of the iceberg.

WHERE DO WE GO FROM HERE?

Faced with such an uncertain future, what should the conservation community do? The question is frighteningly difficult. To my knowledge, no major conservation organization has seriously assessed the future in this way. Yet if the conservation movement fails to ask itself the hard questions, scarce resources will be frittered away in promoting lost causes.

Recognizing that there is not enough money to go around, conservation organizations must invest wisely. Money spent on training, equipment, or policy reform will have been spent in vain if a country later self-destructs in a convulsion of dictatorship or civil war. I would also question the value of investing in countries in which journalists are assassinated for exposing official corruption, for transparency is required if parks are not to be lost to official neglect and corruption.[25]

We must acknowledge the grim reality that biodiversity conservation will fail altogether in some countries through the combined effects of weak institutions, corruption, and social instability. As the global balance between population and resources deteriorates even further, it will become increasingly evident which countries those are. But if organizations wait until all is lost before redirecting spending, promising opportunities will have been missed in the interim. It is thus time to look around the world's tropical regions and to consider triage before the hard decisions are preempted by history. The following are my considered appraisals for the major tropical regions.

West Africa

Nature has all but disappeared in West Africa, and the necessary social order and discipline to conserve what little remains are lacking. Rather than pour money into conservation efforts that are doomed to fail, I would maintain the most spectacular endemic species, such as the pygmy hippopotamus and the diana monkey, in captivity as reminders of our collective impotence to hold back the forces of extinction.

Central Africa

Although much of central Africa's primary forest remains, most of its regional governments are controlled by corrupt dictators. Given the political uncertainty of the region, I would adopt a wait-and-see policy toward countries such as Gabon and the Democratic Republic of Congo (formerly Zaire). Those that promise to hold on to their forest cover for a few more decades deserve support.

Madagascar

Because Madagascar represents an unrivaled evolutionary museum and a treasure trove of endemism, the island is far too important biologically to be dismissed. Although its environmental situation is desperate, Madagascar has a functional government that is capable of making things happen on the ground, and so it should be viewed as a calculated risk.

Southeast Asia

The desperation of the conservation situation in Southeast Asia is rivaled only by that in West Africa. Domestic pressure to protect the region's biodiversity is virtually nonexistent. A ubiquitous Chinese minority creates a market for everything that breathes, from tigers, high on the food chain, down to turtles and snakes. Rapidly growing economies with insatiable appetites for natural resources bode poorly for the survival of forests. Moreover, with few exceptions, governments in this region exercise top-down authority and are rife with corruption. Overpopulated Vietnam and the Philippines are already beyond the point of no return. Malaysia and Indonesia are perhaps the only strong hopes for the entire region.

Papua New Guinea

Rich in diversity and endemism, Papua New Guinea is a global treasure, on a par with Madagascar. However, it presents exceptionally difficult challenges for conservation. Although 100 percent of the land belongs in principle to tribal councils, the Forest Authority of the central government ignores this inconvenient technicality. In 1996, for example, it drew up a National Forest Plan consigning more than half the country's forests to logging. According to an IUCN newsletter, "Forestry areas are declared over existing conservation areas; proposed conservation areas have been disregarded; and regions identified as having the highest priority for biodiversity conservation are slated for logging."[26] Social and legal conditions in Papua New Guinea must change dramatically before this country sees any real conservation progress.

Central America

In Central America, Mexico stands out for its extraordinary ecological variety, high endemism, and rich biodiversity. It harbors more reptiles—more than 1,000 species—than any other country. Unfortunately, Mexico's commitment to its parks can charitably be described as weak. Moreover, the country is already overcrowded at 85 million, and its population is expected to balloon to 150 million by 2030.[27] The society is burdened by corruption from top to bottom and an entrenched power structure.[28] Therefore, as a conservation prospect,

Mexico is disappointing. Guatemala is a social tinderbox. Nature has already been extinguished in El Salvador; Honduras, Nicaragua, and Panama offer modest potential. Costa Rica, with the world's most forward-looking government, magnificent parks, and a quarter of its national territory under protection, is a shining example to the rest of the world and the best bet in the region.

South America

Here lie the best prospects for conservation in the Tropics outside of Costa Rica. South America is the least despoiled continent, possessing the largest area of remaining primary habitat. Population densities are still low to moderate, and though the population continues to expand, growth rates are declining in every country. The record of economic development has been spotty, with too much emphasis on the export of natural resources. However, recently instituted macroeconomic reforms have improved economic prospects for the future. Nearly every country now supports vibrant conservation organizations. The greatest challenges lie in overcoming corruption and the frontier mentality that pervades the entire continent.

Taking the Tropics as a whole, the best prospects for achieving the institutional strength and stability required to protect biodiversity over the long run exist in a minority of countries, in which most or all of the following conditions prevail: low to moderate population density, falling birth rates, improving education, rising standards of living, increasing urbanization, a dwindling rural population, freedom of speech and information, and, above all, democracy. Countries meeting many or all of these criteria have a chance of achieving stabilization before the last tree falls; those that cannot meet many of these criteria should be given lower priority in any attempt at triage.

That said, the prospects for conserving biodiversity should not be viewed as universally hopeless. The world is experiencing a profound revolution in attitudes and values. Dictatorships are yielding to democracy in Latin America, Africa, and Asia. In scores of countries, women and members of ethnic minority groups are gaining rights they never had before. Television and the Internet are opening a window for information exchange in societies hitherto kept in the dark by state-controlled media. Such trends are undeniably positive. What I find most exasperating is the nearly unanimous reluctance of politicians in both developed and developing countries to promote more enlightened policies toward natural resources and biodiversity. All it might take to carry the world over this next great hurdle is one inspired leader who would resist the environmental despoilers and make people understand that in the end, quality of life matters more than the bottom line.

12

NATURE,
A GLOBAL
COMMONS

THE MUCH ACCLAIMED Brundtland report, the World Commission on Environment and Development's *Our Common Future*, anticipates a worldwide doubling of population and quintupling of economic activity before the numbers level off.[1] When I try to contemplate a world with five times today's level of economic activity, my imagination fails me. Five times the energy consumption. Five times the pollution and waste. Five times the demand on natural resources. Even with improved technologies, I wonder whether such a frenzied global economy is possible, much less sustainable. But of one thing I am sure. It is not a world I would want to live in.

Optimally, a world I would want to live in would have far fewer people than are alive now. It would be a world in which resources were not stretched to the limit. Rivers would reach the sea, even in the dry season. Renewable energy derived from hydroelectric, wind, and solar power would supply humanity's needs. The air would be good to breathe and the water safe to drink, free of pesticides, heavy metals, industrial toxins, and human waste. The oceans would once more abound with fish and coastal waters with crabs, shrimp, oysters, and clams. Living well below natural limits would ensure that resources of all kinds were plentiful and cheap. Even the poor could afford to eat well.

A world I would want to live in would not be monopolized, or even dominated, by humans. Much of the landscape would remain wild. Nature would prevail except in small enclaves where favorable environmental circumstances made life especially congenial for humans. Most people would live in small communities, where social constraints and a lack of anonymity would work to keep crime rates low. Amid abundant resources and without the stresses of growth, nations would have less reason to go to war. Production would be managed by automated technologies, leaving people ample leisure time in which to pursue the arts and to enjoy nature.

Am I a misanthrope? I don't think so. In fact, I truly believe that many people would find this vision of the future, or something like it, preferable to the lives they live now. Yet even in the most democratic societies, people fail to communicate their visions of a better future to politicians. In turn, politicians, being fixated on the present, do not articulate such visions either. Instead, they repeat the mantra of growth while promoting the illusion that more is better. It is an expedient way of avoiding responsibility, for guiding people toward a truly sustainable future requires a quality of leadership the world has rarely seen. Yet enlightened leadership seldom springs from an electoral process that resembles, as it does today, a popularity contest more than a contest of ideas.

In a free and open society, reason and objectivity can be drowned out in a cacophony of competing opinions. Amid the clamor, special interests often conspire to override popular opinion. The public can be fooled by cynical misinformation campaigns, such as the one being perpetrated by the so-called wise use movement, which promotes economic development over environmental protection.[2] Imperfect as it may be as a political system, modern popular democracy provides even the lowliest citizen with a voice in governance. When public opinion coalesces, even well-financed opposition must give way to the majority. In the United States, increasing awareness has created a large majority that favors protection of the environment, including preservation of biodiversity and endangered species. Consequently, I am confident that objectivity and popular opinion will eventually prevail in the United States to bring conservation and development into balance.

I wish I could be as confident about the prospects for conservation in the developing world. Successful implementation of tropical parks faces challenges on many fronts: cultural, social, economic, legal, and political. What effective conservation requires above all else is institutional strength and durability. Many countries currently lack robust institutions, so ways must be found to strengthen them. But institution building is a long-term process, whereas the need to protect nature is immediate.

To put the matter in perspective, one need only compare the nature of institutions in, say, northern Europe with those in a hypothetical tropical country. Northern Europe is a region with strong institutions. Its countries are inhabited by homogeneous populations with deep, shared historical roots. These countries have achieved prosperity and population stability in open, democratic societies distinguished by the world's most equitable income distributions.[3] Educational levels are among the highest anywhere. The work ethic is accepted by all levels of society, including the rich. A broad sense of collective social responsibility pervades the culture, resulting in the highest tax rates in the world. Nevertheless, taxes are paid by rich and poor alike with relatively few irregularities. Levels of corruption are the lowest in the world.[4] Social and economic interactions are regulated by law, and the law is fairly and openly administered in courts, which are largely free of political influence. Civilians hold power over military commanders, and military personnel are excluded by law from involvement in political activity. Elections are held on a regular schedule, and changes of government occur routinely and without incident.

Now consider a typical tropical country, whether in Africa, Asia, or Latin America. To be sure, conditions differ across countries and continents, but certain cultural attributes are common to many tropical developing countries. Among the common denominators throughout the tropical world are grossly skewed distributions of power, education, and income. Systematic corruption and abuse of power are pervasive and lead to rigidly hierarchical social organization and a lack of upward mobility. National populations are rarely homogeneous; more commonly, there are dominant and subordinate ethnic groups with distinct cultures and languages. Competition between ethnic groups can be intense; if not, a shaky stability persists because subordinate groups are systematically suppressed by the dominant group, resulting in simmering resentment.

Competition between ethnic groups diminishes any sense of shared national destiny or collective social responsibility. Social welfare programs are consequently meager or nonexistent. Tax collection is inefficient and rife with corruption. The rich rarely pay their due. Fewer than 1 in 1,000 Pakistanis, for example, are registered taxpayers.[5] Because of inefficient tax collection, government revenues derive mainly from regressive measures such as sales taxes and levees on exports and imports. Laws exist on the books, but many are capriciously enforced. Police and the courts are corrupt, for sale to the highest bidder.

Top-down administration of power stifles individual initiative and bottom-up participation in political and administrative processes. In a great many tropical countries, superficially democratic or not, the military wields ultimate power. Military commanders can thus abuse the law with impunity. Elections, if any, are held sporadically, and the outcomes are often manipulated by the party in power.

A culture of corruption begets a general disregard for the law. Lawlessness at the top trickles down to lawlessness at all levels of society. A corrupt, lawless society is incapable of the resolute dedication required to manage resources wisely and to protect parks from illegal incursions.

Institutional weaknesses are particularly pronounced in Africa. The point is passionately made in a letter to the *Economist* by George Ayittey, president of the Free Africa Foundation, from which I quote:

> Sir—As a black African, I find your treatise on "Why poor nations remain poor" (May 25th) missing a key factor: the nature of government itself. . . . In most third-world countries, the state itself has been hijacked by gangsters, with every key institution (judiciary, banking, military, press, etc.) debauched. Sub-Saharan Africa—the poorest region in the third world—is a prime example. Brutally inefficient and grotesquely incompetent authoritarian regimes proliferate. The underlying ethics of authoritarianism are self-aggrandizement and self-perpetuation. Access to political power guarantees fabulous wealth. (The richest people in Africa are heads of state and ministers.) Thus, the political and economic systems are fused: it is futile to reform one without the other. What keeps Africans poor is their powerlessness to remove predatory governments or force existing ones to adopt the right policies in a peaceful way.[6]

When the factor of social institutions is brought into the picture, the challenge of conserving tropical biodiversity comes into sharp focus. Corrupt, predatory governments designed to favor the few at the expense of the many are the norm in parts of the tropical world. Laws can be mere window dressing, offering a facade behind which the dominant classes enrich themselves, frequently through unimpeded access to natural resources. Some regimes lack even the most rudimentary social justice or redress for the poor, perpetuating a system in which the only enforced law is "might makes right." And in these countries, the poor must do whatever they can to survive, even if survival means ravaging irreplaceable resources.

How can parks be protected and biodiversity preserved in such dysfunctional societies? Parks cannot be maintained without order and discipline, but order and discipline are antithetical to many of the corrupt, freewheeling societies of the developing world.

These facts of life, unpalatable as they may be, fuel my concern that nature will not survive the twenty-first century in large parts of the tropical world. If the international conservation organizations continue with business as usual, pouring their funds into integrated conservation and development projects (ICDPs), nature will be the loser.

What, then, are some alternatives that might put nature protection on a sounder footing? Before addressing the question, I would like to consider the question of who will pay for nature protection in the developing world.

The West currently foots the bill for much of what passes as nature protection in tropical countries. The West pays because preservation of nature is a Western cultural value. Fund-raising campaigns based on appeals to save the tiger, elephant, rhino, or what-have-you bring in millions of dollars, the vast majority of which emanate from North America, Europe, Australia, and New Zealand. The rest of the world either is too poor to make contributions or does not care.

Caring can be measured through a concept known as existence value. The term refers to nonmarketable commodities that people value for their intrinsic qualities. Many U.S. citizens happily pay taxes to support museums and other cultural institutions they may never visit. They receive their money's worth simply by knowing that great art or historical collections are serving to educate the younger generation and are preserved for posterity.

Tigers and rhinoceroses have high existence value. That is why they and other "charismatic megavertebrates" are chosen as the focal points of fund-raising campaigns, despite the fact that few potential donors are likely ever to see them in the wild. People are happy simply to know that these animals are there, that they exist.

The same argument can be applied to the parks that shelter charismatic fauna. But now I shall add a slight twist. Existence values are not fixed. Like other kinds of personal or cultural values, they are relative and are subject to change with changing circumstances.

Two of my Duke University colleagues, Randy Kramer and Priya Shyamsundar, recently studied the existence value of a park in Madagascar.[7] Villagers living near the boundary were asked what the park meant to them. The responses were unequivocal. The park was an arbitrary imposition of the government and infringed on their use of the land. Before the park existed, the villagers had lived without restrictions. Now they were prohibited from gathering fuelwood, hunting, cutting building materials, or clearing farm plots in the forest. Potential benefits to be derived from the park, such as a supply of potable water or future ecotourism revenues, were discounted by the villagers as too abstract and remote. To the villagers, therefore, the park had a negative existence value.

But local villagers are not the park's only constituency. Malagasies queried in the capital city gave responses that were neutral or mildly positive, although few had any real notion of what a park is or what purpose it serves. Similar stories could be told for many other developing countries. On balance, it can be concluded that parks in poor tropical countries have, at best, a zero existence value. It is little wonder, then, that park budgets are so woefully inadequate in so many countries.

Many of the big, beautiful, and exciting animals westerners value live in countries where they are not valued or are valued only for consumptive purposes. If the citizens of those countries decline to protect their own flora and fauna, then, by default, the people who do value them—we residents of the developed world—will have to pay to protect them, or else no one will. There is no escaping the logic.

Relative to the scale of effort required, the amount residents of the developed world voluntarily contribute to save tropical nature represents a drop in the bucket. Discounting pass-through funds derived from USAID or the international banks, all U.S. conservation organizations together spend less than $100 million annually on international programs. Europeans contribute a similar amount. The thought that hundreds of parks in scores of countries could be supported and protected for $200 million per year is absurd. To put that amount into perspective, the 1997 budget of the National Park Service in the United States was $1.3 billion.[8] Tropical nature simply cannot be saved unless new funding mechanisms are created.

One such mechanism, the debt-for-nature swap, created a flurry of excitement in the conservation community in the late 1980s, when it was first proposed. A brilliant concept, it involved the idea that a government would buy back debt obligation at a discount, with the proceeds of the sale deposited in an account earmarked for conservation. A few such deals were consummated, but many others languished, unsigned, on the desks of finance ministers. Many countries were seeking absolution of their debts or were hoping that the value of their obligations would sink to zero on financial markets so they would no longer represent liabilities. Debt-for-nature accords put a definite value on outstanding debt, setting a level from which there was no easy retreat. Indebted governments have lately shied away from new debt swap agreements, and the whole idea appears to have run out of steam.

In 1990, another funding mechanism came into being, generating a new wave of excitement. Launched by the World Bank, the new program was called the Global Environment Facility (GEF). The bank's directors announced that sizable appropriations would go into the GEF to support environmental programs, including programs to strengthen parks.

To date, the hoped-for conservation bonanza has yet to materialize, although some worthy projects have been funded. In practice, the allocation of GEF resources is highly political. The funds go mainly to governments, and they must pass through multiple bureaucratic layers in a process so byzantine that final approval of a grant routinely takes two years or longer. Unless the GEF acquires new life and vigor in coming years, it promises to be just another flash in the pan.[9]

A more traditional mechanism for funding conservation projects is foreign aid. Like the GEF, foreign aid can potentially contribute billions to conservation, an order of magnitude more than the $200 million or so that private contributions can provide. As a longtime observer in Perú, I have encountered a wide range of foreign assistance programs. Of the half dozen programs with which I am familiar, USAID's is one of the least effective.

USAID has become a whipping boy in the trench warfare of Washington politics, but not for its ineffectiveness. At issue is the pinched national budget and the growing unpopularity of foreign aid itself, a phenomenon known as donor fatigue.[10] The unpopularity of foreign aid reflects more the machinations of politicians than a lack of generosity on the part of the people of the United States. The U.S. public has been grossly deceived by misinformation issuing from political organizations that are implacably hostile to foreign aid in any form. A poll published in the *New York Times* reported 15 percent as the median figure offered by respondents when asked what fraction of the federal budget was spent on foreign aid. When asked whether this was "too much," "about right," or "not enough," those interviewed split into three camps, with the "about rights" having the plurality.[11]

Apparently, few U.S. citizens realize that theirs is one of the stingiest of the developed countries, spending less than 1 percent of the federal budget on foreign aid. Moreover, in 1990, more than half of the USAID budget went to just four countries, Israel, Egypt, Panama, and the Philippines.[12] At least three of these are needy countries, but politics clearly holds precedence over altruism in the allocation of USAID funds.

The negative propaganda regarding foreign aid is so pervasive that when U.S. residents were asked, in a different poll, what programs should be cut to help balance the federal budget, foreign aid topped the list. Citizens who are bombarded with innuendo and misinformation cannot make informed decisions. We have legislated truth in advertising, but there is no equivalent compulsion to respect veracity in politics or on talk radio.

Thus, USAID, potentially a major source of funds for biodiversity, is being squeezed into insignificance by a hostile Congress and an ignorant public. USAID needs to be reformed, not extinguished. If the United States retreats from its long-standing role as global leader, the job of saving biodiversity will increasingly pass to the Europeans and the Japanese.[13]

But political climates are fickle and subject to change. So let us suppose that the current U.S. retreat from foreign assistance is only a temporary aberration, a swing of the political pendulum. Let us further suppose that the United States and the other leading developed countries could agree on a regular allocation for global biodiversity protection so that billions of dollars, rather than millions, could annually flow into parks and park protection. What then?

A good starting point would be to establish national conservation trust funds, essentially endowments in support of parks. Several have been created to date (for example, in Bhutan and Mexico), in an en-

couraging trend that I hope will not be just another passing fad. Conservation trust funds can provide the financial stability that is essential
to maintaining long-term programs and to attracting able guards and
administrative personnel to serve in park protection.

Trust funds offer an enormous advantage over short-term projects
in providing needed financial stability and predictability, but they are
not without problems. Where are the funds to be invested? Who is to
administer them? How will the revenues be disbursed within the recipient countries? What bodies will ultimately be responsible for nature
protection? And, as is most likely, if the funds are to be established
by international organizations, what role will national governments
have? All these are potential sticking points in negotiations between
international bodies and national governments.

Certainly, all national governments jealously guard their authority
and prerogatives. The Peruvian government, for example, although
unable at times to pay park guards, would be extremely reluctant to
cede responsibility for nature protection to a foreign entity. Finding
mechanisms through which the developed countries can make significant, sustained, and direct contributions to nature protection will require a lot of creative thinking and delicate negotiation.

Adequate financing in itself does not guarantee successful parks if
enforcement is not rigorously implemented. Enforcement is the sine
qua non of biodiversity preservation. But the enforcement issue
touches even deeper nationalistic nerves than does the issue of financing mechanisms, so it presents a much greater challenge.

Still, a wide range of options for enhancing enforcement exists, suggesting that the challenges may not be insurmountable. The decision
of which option would serve best in a given instance would depend
on the strength of the country's institutions and the receptivity of its
government. Some options rely more heavily on national institutions;
others, on international institutions.

Even poor countries can protect parks without external help if the circumstances are positive. The Royal Chitwan National Park in Nepal, described in chapter 5, is a shining example. There, 800 soldiers of the Royal Nepalese Army guard the largest extant population of one-horned rhinoceroses. Chitwan, however, is a special case on two counts. First, gate receipts pay much of the cost of stationing the troops in and around the park. Second, corruption is not a major problem because the park is a pet project of Nepal's king.

The Chitwan model could not be widely applied because very few parks in the world attract enough visitors to support an army on entrance fees. But neither would 800 troops be required to guard most parks, since few parks contain resources as valuable as rhinoceroses.

In a number of countries, military forces have been deployed for park protection when all else has failed, but seldom as a permanent guard force. Assignment to long-term duty would require the approval of both the government in power and the military high command, but the assignment of military forces is vulnerable to all the vicissitudes of national politics. The ultimate goal should be to uncouple the business of nature protection from national politics.

In my opinion, the best—perhaps the only—hope lies in the internationalization of nature protection, a measure for which there is compelling logic. Biodiversity transcends national boundaries and belongs to no one. The notion that biodiversity is a global commons, belonging only to the planet earth, is not new. To date, however, governments and international bodies have failed to take the idea seriously. Yet whatever the philosophical merits of regarding biodiversity as a global responsibility, more than philosophy will be needed to impress the heads of state who will be asked to compromise their nations' sovereignty in the interest of nature protection.

So many environmental problems transcend international boundaries that enforcement mechanisms will eventually be necessary to ensure the compliance of individual countries with international standards and agreements. Many countries are not yet ready to concede that they are part of a larger global community, but there are exceptions. I quote from Mario Boza, who at the time of this statement was vice minister of natural resources, energy, and mines in the government of Costa Rica and who is the distinguished past director of Costa Rica's national park system:

> Many organizations use funds to maintain an international bureaucracy rather than supporting direct conservation in the field. We do not need more planning studies and documents to tell us what to do, but instead we need funds to make environmental conservation a reality at the grass-roots level. International environmental standards should be set by a United Nations environmental organization that is empowered to infringe on the sovereignty of individual states in environmental matters.[14]

By suggesting that the United Nations be given authority to override the sovereignty of member nations, Vice Minister Boza makes a longer leap into the future than I am suggesting here. It will be some time before the world at large allows international agencies to impose enforcement of environmental standards on unwilling governments. Until such a time, governments might lend their voluntary support to less radical measures. After all, internationalization of nature protection does not have to come at one leap; it can be introduced in incremental steps, each entailing only minor concessions to national sovereignty.

A first step might be to create internationally financed elite forces within countries, counterparts of the rangers who protect national parks in the United States and are legally authorized to carry arms and make arrests. In Washington, D.C., where a significant fraction of the city is federal parkland, the National Park Service maintains its own courts and judges.

An internationally financed elite corps, well trained and well paid, could attract highly qualified citizens into the service of biodiversity conservation. Good pay and independence from national and local political influence would diminish conflicts of interest and incentives for corruption. Whether such a park protection corps would require its own courts, judges, and jails would depend on conditions in the recipient countries and the details of the underlying international agreements.

Although park personnel in a number of countries are currently financed from abroad, the next step could be to give the park protection service independence from local police and courts, enabling it to implement the enforcement process from start to finish. The reactions of national governments to such a suggestion are certain to be mixed. At worst, the idea of an autonomous enforcement authority would be rejected as an unacceptable intrusion into a country's internal affairs. At best, some governments might be relieved to have the politically unpopular enforcement function taken out of their hands.

Of course, one of the largest stumbling blocks in elevating park protection to an elite service with a great deal of autonomy would be the opposition of police and military forces, always jealous of their authority. How much better the world could be if more countries would follow the lead of Costa Rica and renounce militarism altogether.

A more direct option is to buy land and protect it. Had the millions of dollars that have gone into unsuccessful ICDPs instead been invested in land acquisition, the donor nations would now have title to huge tracts of tropical forest. Already, both organizations and individuals have purchased habitat for nature preserves in Costa Rica, Brazil, Chile, and doubtless in other countries.[15]

Having the appeal of directness and simplicity, the purchase of wildlands in the Tropics is nevertheless a limited option. Many governments do not allow foreigners to buy land. Where foreigners are prohibited from landownership, titles could be held by local conservation organizations. But in the long run, titles are inconsequential if governments fail to honor them. In much of Latin America, local police forces and courts resist taking action against squatters and other opportunists and titles are actively honored only if the land is in productive use, which means land that has been cleared.

Beyond the immediate difficulty of protecting purchased wildlands from encroachment lies a long-term risk of expropriation. One hopes this will be a receding risk, given the rapidly growing interdependence of national economies. Facing the prospect of international trade sanctions, countries will have more to lose than to gain from expropriating foreign-owned properties.

The most radical option would be full internationalization of nature protection under the auspices of the United Nations or some other yet to be constituted international body. If peacekeeping has been widely accepted as an international function, why not nature keeping? Peacekeeping has become a routine mechanism for maintaining order in countries unable to do so by themselves. If local park guards are too weak or too subject to corruption and political influence to carry out their duties effectively, internationally sponsored guards could be called in to help. As foreigners, they would be independent of local pressures and thus better able to exercise authority.

In some countries, international nature keeping might have to be sustained for long periods, much as peacekeeping forces have been maintained in Cyprus. In others, external intervention could be short-term. In countries where corruption and lawlessness are pervasive, armed troops with full enforcement authority would be required. Elsewhere, unarmed personnel might engender sufficient respect to deter illegal activity.

As an adjunct to international nature keeping, there could be a

Nature Corps, analogous to the Peace Corps in the United States. A Nature Corps could serve in many capacities, acting as a watchdog for parks, reinforcing local guards, assisting in administration, conducting environmental education programs, and so forth. The presence of outsiders having direct connections to higher levels of authority could deter petty opportunists, present in every society, who freely engage in illegal activity when the risks are perceived as low. Given the extraordinary level of concern for the future I see among young people, I am convinced that there would be no shortage of volunteers who would happily dedicate a year or two of their lives to helping nature survive.

Finally, there needs to be a global watchdog organization, somewhat analogous to Amnesty International, to monitor the health of parks around the world. Without such an organization, the health of parks will never be a matter of public record. Today, hundreds of paper parks in scores of countries are known to only a relative handful of people, and the current status of many is undocumented. The World Conservation Monitoring Centre (WCMC) in Cambridge, England, does a capital job of compiling statistics on protected areas, but many of the data issue from government sources. Even when governments issue accurate information—and in many cases they do not—they may be reluctant to reveal how many squatters have invaded a park, how much of a park's area has been compromised by illegal logging, or how many miners are deep in the interior panning gold. Either the scope of the WCMC should be expanded or a new organization should be created to fill the need.

Although parks can be monitored from space, satellite images reveal only certain classes of violations, such as clear-cut logging and wholesale land clearing. Today's space technology cannot detect small-scale violations such as poaching, fuelwood gathering, selective logging, and clearing of single-family garden plots. For the detection of these more subtle, but sometimes equally insidious, forms of degradation, there is no substitute for on-the-spot reconnaissance. With accurate eyewitness reporting on the status of parks and effective dissemination of the information, knowledge can be power, since no government wants to have its failings illuminated in a public spotlight.

I have suggested steps that can be taken to put nature protection on a sound footing, not only in the Tropics but around the world. Putting into practice some of these suggestions, or others that could be imagined, will require creativity, dedication, and skillful diplomacy. But whatever route is taken, the ultimate goal must be to protect nature from the forces that threaten to destroy it.

More broadly, the task of bringing sustainable development to the world will be the most daunting challenge ever faced by humanity. The bottom-up approach currently being pursued by many international organizations in the guise of ICDPs is tantamount to social engineering. In the United States, government-sponsored experiments in social engineering have expended hundreds of billions of dollars in funds and yet have achieved only modest progress toward the goal of reducing poverty. The social problems of the big cities remain stubbornly intractable. Attempting to carry out social engineering in a foreign culture that is undergoing rapid social and economic change within the period of a five-year assistance project can, in my view, be likened to pouring water into the Sahara Desert.

Moreover, the bottom-up approach is unlikely to generate lasting change because people near the bottom of the economic ladder are so fundamentally dependent on decisions made at the top. The possibilities open to a farmer, for example, depend on the market prices of the commodities he or she produces as balanced against the costs of water, energy, fertilizer, and transportation. In many developing countries, the prices of agricultural products, as well as those of energy, transportation, fertilizer, and water, are all directly or indirectly set by the government. Changes in any of these prices, made through a top-down process, can decisively alter the economic equations that govern the farmer's decision making.

Whether a farmer decides, for example, to expand the area of land under cultivation by clearing additional forest or to retain the forest for the game and other resources it yields will depend on the economic merits of the alternatives. With the progressive lowering of barriers to international trade, the economic forces to which a farmer responds are rapidly becoming global instead of local or national, as they were in the past. Thus, to an increasing degree, the fate of the tropical forests is held hostage by forces that small players on the ground are powerless to influence.

Decisions made at high levels also drive the tropical timber industry. Timber cannot be extracted from tropical forests on an industrial scale in the absence of infrastructure such as roads and ports. Ultimately, taxpayers pay for infrastructure, thereby unwittingly subsidizing the liquidation of timber resources held in the public domain. This occurs in the United States when the costs of road building and management exceed the price brought by the sale of timber from national forests.

Government policies determine profit margins in the timber industry in other ways as well, especially through the pricing of concessions and the stump fees (royalties) charged for harvested logs. In many tropical countries, concessions are granted as political favors and stump fees are set at rates that lead to high profits while costing governments billions in forgone revenues.[16]

In short, government policies in scores of countries promote forest destruction by providing infrastructure at no cost to industry and by offering financial incentives in the form of subsidies, tax breaks, and even outright giveaways. To my way of thinking, it makes little sense to talk about creating village-level forestry cooperatives when the big players are wheeling and dealing in billion-dollar contracts signed at the level of ministers and presidents. It would make sense to talk about village-level forestry cooperatives if national policy actively favored them, but in too many countries that is not the case.

The big challenges facing conservation, as I see them, are not at the grassroots level but at the level of government policy. Policies affect the use of most renewable resources and set the stage for biological conservation. The areas of policy I consider most crucial to conservation and sustainable development are land tenure, land-use planning, trade, subsidies, taxes, and, finally, investment in parks and biodiversity conservation.

First, who should own the forests? Large-scale privatization of forests, as in Brazil, creates almost insurmountable obstacles for conservation because private land must be purchased to be protected. On the other hand, maintenance of forestlands under state ownership leaves open such options as establishing reserves for indigenous people, instituting adaptive management of quasi-natural forests, and creating future nature preserves.

The greatest danger of retaining forestlands in the public sector is that corrupt or unscrupulous governments might squander a country's entire forest estate in timber concessions, as has already happened in several cases. Despite such risks, I believe that on balance it is better to maintain forests under public ownership because doing so preserves more options for the future.

Countries that retain large tracts of public land will unavoidably be drawn into land-use planning. A national land-use plan can employ optimality criteria to determine the most appropriate uses for each tract of land, and it can require constancy of land use over the long term, as is mandated by legislation governing federal lands in the United States. To be avoided are regulations that permit land use to flip-flop with every economic turn.[17] Ultimately, it is imperative to halt the land-use cascade so that the next generation does not inherit an exhausted planet.[18]

Whether wildlands will continue to exist at the end of the twenty-first century depends more than anything else on government policies. The greatest destroyers of wildlands are major government-sponsored projects. The Transamazonian Highway system and Tucuruí Dam in Brazil, and the Trans-Gabon Railroad are prime examples. An increasing reluctance on the part of the World Bank to sponsor environmentally damaging projects is an encouraging bit of progress. But ultimately, it will be governments, not international organizations such as the World Bank, that will decide whether or not to build roads into the last tracts of tropical wilderness. Roads are the death knell of the rain forest. Wilderness cannot be preserved without restraining road building.

The free-for-all that is unleashed by unregulated access to state lands has become a glaring anachronism, a lingering vestige of an outmoded vision of limitless resources. The era of open access must come to an end, and the sooner the better. Land tenure will have to be defined and enforced the world over, as it has been in the developed countries for decades and even centuries. In the absence of well-defined land tenure and orderly procedures governing land transfers, the "might makes right" principle concentrates landownership in the hands of the powerful. And that, as history amply affirms, is the stuff of revolution.

Trade agreements, subsidies, and taxes are all powerful economic levers that are matters of high-level policy. Subsidies promote the unsustainable use of land, forests, grasslands, and water the world around.[19] Tax breaks accomplish many of the same ends.[20] Some fis-

cal policy is particularly pernicious, examples being resource depletion allowances and the practice in some developing countries of tax-
ing "undeveloped" (still-forested) land at higher rates than cleared
land. A more blatant way of promoting deforestation could hardly be
devised.

Finally, the creation of parks is a quintessentially top-down function, as is the establishment of policies that determine the staffing levels of national park services and the salaries and work conditions of
park personnel. In too many countries, park personnel are drawn
from the lowest rungs of society and receive little or no training or institutional support. Comparisons with the highly educated and thoroughly professional rangers of the National Park Service in the United
States underscore the need for improvement. In dozens of countries,
interagency rivalries that deny the right of park guards to carry arms
and make arrests persist as a major impediment to creating viable
parks.

Numerous policies of nearly every government on earth are antithetical to sustainable development and, ultimately, to conservation
of biodiversity. The need for policy reform is therefore obvious and
urgent, but it is not the only critical issue. In many areas of resource
use but particularly in the use of forests, there is an urgent need to
combat collusion between avaricious business interests and corrupt
or otherwise venal politicians and their military cronies. Such collusions are particularly strong and odious in U.S. politics, and they are a
fact of life in every developing country I know of except, perhaps,
Costa Rica.[21] Breaking these pernicious collusions will require no less
than a redesigning of democracy.

The ultimate challenge, of which halting the rapacious exploitation of forests is only an example, is counteracting the tragedy of the
commons. Experience teaches that restraint in the use of renewable
resources will not spring up from the bottom. Farmers knowingly

mine the soil and irrigate their fields with fossil water, depleting underground aquifers. The timber industry donates vast sums of money to the campaigns of timber-state congressional representatives, who then vote to cut forests in the public domain faster than they are growing.[22] Fishermen lobby against harvest quotas and environmentally soft technologies such as turtle exclusion devices out of fear that their incomes might be compromised.[23]

In many democracies, the selfish demands of the clamorous few are promoted over the diffuse interests of a passive majority. This "tyranny of the minority" is a widely acknowledged failing of democracy, but concrete suggestions for rectifying it hardly enter the political debate. Somehow, we are going to have to turn the corner because humanity cannot forever avert its face from the fact that rational and restrained use of renewable resources offers the only route to future peace and prosperity. The person who leads the way to ending the tragedy of the commons will truly be the person who saves the world.

NOTES

Chapter 2

1. A hectare is equivalent to 2.47 acres.

2. B. Drayton and R. B. Primack, "Plant Species Lost in an Isolated Conservation Area in Metropolitan Boston from 1894 to 1993," *Conservation Biology* 10 (1996): 30–39; C. K. Yoon, "Plant Census Raises the Alarm and Leads to Restoration Effort," *New York Times,* February 13, 1996, p. C4.

3. P. G. Crawshaw and H. B. Quigley, "Jaguar Spacing, Activity, and Habitat Use in a Seasonally Flooded Environment in Brazil," *Journal of the Zoological Society of London* 223 (1991): 357–370.

4. J. Diamond, "Playing Dice with Megadeath," *Discover,* April 1990, pp. 55–59; D. W. Steadman, "Prehistoric Extinctions of Pacific Island Birds: Biodiversity Meets Zooarchaeology," *Science* 267 (1995): 1123–1130.

5. Rockefeller Foundation, *High Stakes: The United States, Global Population, and Our Common Future* (New York: Rockefeller Foundation, 1997).

6. IUCN and WCMC (World Conservation Union and World Conservation Monitoring Centre), *Protected Areas of the World: A Review of Natural Systems* (Cambridge, England: IUCN and WCMC, 1992).

7. A. P. Dobson, A. D. Bradshaw, and A. J. M. Baker, "Hopes for the Future: Restoration Ecology and Conservation Biology," *Science* 277 (1997): 515–522.

8. J. Terborgh and C. P. van Schaik, "Minimizing Species Loss: The Imperative of Protection," in *Last Stand: Protected Areas and the Defense of Trop-*

ical Biodiversity, ed. R. Kramer, C. P. van Schaik, and J. Johnson (New York: Oxford University Press, 1997), pp. 15–35.

9. M. McRae, "Road Kill in Cameroon," *Natural History* 106, no. 2 (1997): 36–75; J. E. Fa, J. Juste, J. Perez del Val, and J. Castroviejo, "Impact of Market Hunting on Mammal Species in Equatorial Guinea," *Conservation Biology* 9 (1995): 1107–1115.

10. "High Hopes Fade in Congo," *Economist,* July 12, 1997, pp. 39–40.

11. "An African for Africa," *Time,* September 1, 1997, pp. 36–40.

12. C. E. G. Tutin and M. Fernandes, "Nationwide Census of Gorilla *(Gorilla g. gorilla)* and Chimpanzee *(Pan t. troglodytes)* Populations in Gabon," *American Journal of Primatology* 6 (1984): 313–336.

13. H. W. French, "An African Forest Harbors Vast Wealth and Peril," *New York Times,* April 3, 1996, p. A3; IUCN and WWF (World Conservation Union and World Wide Fund for Nature), "Transnational Loggers Threaten Africa's Forests," *Arborvitae: The IUCN/WWF Forest Conservation Newsletter* 4 (1996): 12.

Chapter 3

1. J. Terborgh, J. W. Fitzpatrick, and L. Emmons, "Annotated Checklist of Bird and Mammal Species of Cocha Cashu Biological Station, Manu National Park, Perú," *Fieldiana: Zoology,* n.s., no. 21 (1984): 1–29.

2. V. Pacheco et al., "List of Mammal Species Known to Occur in Manu Biosphere Reserve, Perú," *Publicaciones del Museo de Historia Natural, Universidad Nacional Mayor de San Marcos,* ser. a, 44 (1993): 1–12.

3. L. B. Rodriguez and J. E. Cadle, "A Preliminary Overview of the Herpetofauna of Cocha Cashu, Manu National Park, Perú," in *Four Neotropical Rainforests,* ed. A. H. Gentry (New Haven, Conn.: Yale University Press, 1990), pp. 410–425.

4. R. B. Foster, "The Floristic Composition of the Rio Manu Floodplain Forest," in Gentry, *Four Neotropical Rainforests,* pp. 99–111.

5. D. E. Wilson and A. Sandoval, eds., *Manu: The Biodiversity of Southeastern Perú* (Washington, D.C.: Smithsonian Institution Press, 1997).

6. Ibid.

7. J. M. Diamond, *Guns, Germs, and Steel: The Fates of Human Societies* (New York: Norton, 1997).

8. Population Reference Bureau, *1996 World Population Data Sheet* (Washington, D.C.: Population Reference Bureau, 1996).

Chapter 4

1. J. G. Robinson and K. H. Redford, *Neotropical Wildlife Use and Conservation* (Chicago: University of Chicago Press, 1991); M. Alvard and H. Kaplan, "Procurement Technology and Prey Mortality among Indigenous Neotropical Hunters," in *Human Predators and Prey Mortality*, ed. M. Stiner (Boulder, Colo.: Westview Press, 1991).

Chapter 5

1. "Factfile," *People and the Planet* 5, no. 4 (1996): 12; IUCN Forest Conservation Programme, *Forest Conservation Programme Newsletter* 9 (1991): 5.
2. P. G. Crawshaw and H. B. Quigley, "Jaguar Spacing, Activity, and Habitat Use in a Seasonally Flooded Environment in Brazil," *Journal of the Zoological Society of London* 223 (1991): 357–370.
3. Eduardo Alvarez, personal communication, 1997.
4. Christof Schenk, personal communication, 1996.
5. R. Barbault and S. D. Sastrapradja, coords., "Generation, Maintenance, and Loss of Biodiversity," in *Global Biodiversity Assessment*, ed. V. H. Heywood (Cambridge, England: United Nations Environment Programme, 1995), pp. 193–274.
6. C. A. Peres and J. W. Terborgh, "Amazonian Nature Reserves: An Analysis of the Defensibility Status of Existing Conservation Units and Design Criteria for the Future," *Conservation Biology* 9 (1995): 34–46.
7. E. Dinerstein and E. D. Wileramanayake, "Beyond 'Hotspots': How to Prioritize Investments to Conserve Biodiversity in the Indo-Pacific Region," *Conservation Biology* 7 (1993): 53–65.
8. Peres and Terborgh, "Amazonian Nature Reserves."
9. Ibid.
10. Ibid.
11. Ibid.
12. "Forests Cleared in Guatemalan Park," *Oryx* 26 (1992): 197.
13. P. Mercer, "Colombia's National Parks Are in a Losing Battle for Survival," *New York Times*, March 28, 1995, p. B11.

14. World Resources Institute, *World Resources 1990–1991* (Washington, D.C.: World Resources Institute, 1990).

15. D. W. Yu, T. Hendrickson, and A. Castillo, "Ecotourism and Conservation in Amazonian Perú: Short-Term and Long-Term Challenges," *Environmental Conservation* 24 (1997): 130–138.

16. IUCN and WWF (World Conservation Union and World Wide Fund for Nature), "Liberia: The Plunder Continues," *Arborvitae: The IUCN/WWF Forest Conservation Newsletter,* August 1997, p. 15.

17. J. Randal, "Evacuation of Relief Workers Worsens Liberia's Woes," *Washington Post,* April 14, 1996, p. A24.

18. C. Martin, *The Rainforests of West Africa: Ecology, Threats, Conservation* (Basel, Switzerland: Birkhäuser Verlag, 1991).

19. Ibid.

20. J. Terborgh, *Tropical Deforestation* (Burlington, N.C.: Carolina Biological Supply Company, 1992).

21. K. H. Redford, "The Empty Forest," *BioScience* 42, no. 6 (1992): 412–422.

22. Population Reference Bureau, *1996 World Population Data Sheet* (Washington, D.C.: Population Reference Bureau, 1996).

23. C. E. G. Tutin and M. Fernandes, "Nationwide Census of Gorilla *(Gorilla g. gorilla)* and Chimpanzee *(Pan t. troglodytes)* Populations in Gabon," *American Journal of Primatology* 6 (1984): 313–336.

24. P. Oberlé, ed., *Madagascar: Un Sanctuaire de la Nature* (Antananarivo, Madagascar: Kintana, 1981).

25. R. Dewar, "Extinctions in Madagascar: The Loss of the Subfossil Fauna," in *Quaternary Extinctions,* ed. P. Martin and R. Klein (Tucson: University of Arizona Press, 1984), pp. 574–593.

26. N. Myers, *Deforestation Rates in Tropical Forests and Their Climatic Implications.* Report prepared for Friends of the Earth Trust, London, 1989.

27. M. Wells and K. Brandon, *People and Parks: Linking Protected Area Management with Local Communities* (Washington, D.C.: World Bank, 1992).

28. Ibid.

29. L. Vigne and E. Martin, "Good News in Nepal," *Wildlife Conservation* 98 (1995): 64.

30. Wells and Brandon, *People and Parks.*

31. Ibid.

32. Ibid.

33. Myers, *Deforestation Rates.*

34. Ibid.

Chapter 6

1. L. L. Loope and S. M. Gon III, "Biological Diversity and Its Loss," in *Conservation Biology in Hawai'i*, ed. C. P. Stone and D. B. Stone (Honolulu: University of Hawaii Press, 1989), pp. 109–117; J. Tuxill and C. Bright, "Losing Strands in the Web of Life," in *State of the World 1998*, ed. L. Starke (New York: Norton, 1998), pp. 41–58.

2. J. M. Diamond, "Human Use of World Resources," *Nature* 328 (1987): 479–480; G. C. Daily and P. R. Ehrlich, "Population, Sustainability, and Carrying Capacity," *BioScience* 42 (1992): 761–771; P. M. Vitousek, H. A. Mooney, J. Lubchenco, and J. M. Mello, "Human Domination of the Earth's Ecosystems," *Science* 277 (1997): 494–499.

3. R. Kramer, C. van Schaik, and J. Johnson, eds., *Last Stand: Protected Areas and the Defense of Tropical Biodiversity* (New York: Oxford University Press, 1997).

4. J. Terborgh and B. Winter, "A Method for Siting Parks and Reserves with Special Reference to Colombia and Ecuador," *Biological Conservation* 27 (1983): 45–58.

5. IUCN (World Conservation Union), *Protected Areas of the World: A Review of National Systems*, vol. 4, *Nearctic and Neotropical* (Gland, Switzerland: IUCN, 1992).

6. D. L. Hawksworth and M. T. Kalin-Arroyo, coords., "Magnitude and Distribution of Biodiversity," in *Global Biodiversity Assessment*, ed. V. H. Heywood (Cambridge, England: United Nations Environment Programme, 1995), pp. 107–191.

7. IUCN and WWF (World Conservation Union and World Wide Fund for Nature), "News from around the World," *Arborvitae: The IUCN/WWF Forest Conservation Newsletter* 6 (1997): 4.

8. International Council for Bird Preservation, *Putting Biodiversity on the Map: Priority Areas for Global Conservation* (Cambridge, England: International Council for Bird Preservation, 1992).

9. R. A. Mittermeier, "Primate Diversity and the Tropical Forest: Case

Studies from Brazil and Madagascar and the Importance of the Mega-diversity Countries," in *Biodiversity,* ed. E. O. Wilson (Washington, D.C.: National Academy Press, 1988), pp. 145–154.

10. WWF-US Conservation Science Program and WWF-Canada, "Terres-trial Ecoregions of the United States and Canada," draft map (Washing-ton, D.C.: World Wildlife Fund, n.d.).

Chapter 7

1. R. Abramson, "Rooting Out Wild Pigs' Mischief in Great Smoky Moun-tains," *Los Angeles Times,* July 21, 1992, p. A5.

2. T. L. C. Casey and J. D. Jacobi, "A New Genus and Species of Bird from the Island of Maui, Hawaii *(Passeriformes: Drepanididae),*" *Occasional Papers of the Bernice Pauahi Bishop Museum* 4 (1974): 133–176; H. D. Prann, P. L. Bruner, and D. G. Berrett, *A Field Guide to the Birds of Hawaii and the Tropical Pacific* (Princeton, N.J.: Princeton University Press, 1987).

3. J. M. Scott, S. Mountainspring, F. L. Ramsey, and C. B. Kepler, *Studies in Avian Biology,* vol. 9, *Forest Bird Communities of the Hawaiian Islands: Their Dynamics, Ecology, and Conservation* (Berkeley, Calif.: Cooper Ornitho-logical Society, 1986).

4. R. T. T. Forman, *Landscape Ecology* (New York: Wiley, 1986).

5. E. Willis, "Populations and Local Extinctions of Birds on Barro Colo-rado Island," *Ecological Monographs* 44 (1974): 153–169.

6. F. M. Chapman, *My Tropical Air Castle: Nature Studies in Panama* (New York: Appleton, 1929).

7. Willis, "Birds on Barro Colorado Island."

8. E. S. Morton, "Reintroducing Recently Extirpated Birds into a Tropical Forest Preserve," in *Endangered Birds: Management Techniques for Preserv-ing Threatened Species,* ed. S. A. Temple (Madison: University of Wiscon-sin Press, 1978), pp. 379–384.

9. B. A. Loiselle and W. G. Hoppes, "Nest Predation in Insular and Main-land Lowland Rainforest in Panama," *Condor* 85 (1983): 93–95.

10. L. H. Emmons, "Comparative Feeding Ecology of Felids in a Neotrop-ical Rainforest," *Behavioral Ecology and Sociobiology* 20 (1987): 271–283.

11. J. Terborgh, "Maintenance of Diversity in Tropical Forests," *Biotropica* 24 (1992): 283–292.

12. J. R. Karr, "Avian Survival Rates and the Extinction Process on Barro Colorado Island, Panama," *Conservation Biology* 4, no. 4 (1990): 391–397; J. R. Karr, "Avian Extinction on Barro Colorado Island, Panama: A Reassessment," *American Naturalist* 119 (1982): 220–239.

13. E. G. Leigh Jr., A. S. Rand, and D. M. Windsor, eds., *The Ecology of a Tropical Forest: Seasonal Rhythms and Long-Term Changes* (Washington, D.C.: Smithsonian Institution Press, 1982).

14. J. G. Robinson, *Seasonal Variation in Use of Time and Space by the Wedge-Capped Capuchin Monkey,* Cebus olivaceus: *Implications for Foraging Theory,* Smithsonian Contributions to Zoology No. 431 (Washington, D.C.: Smithsonian Institution Press, 1986).

15. J. Terborgh, L. Lopez, and J. S. Tello, "Bird Communities in Transition: The Lago Guri Islands," *Ecology* 78, no. 5 (1997): 1494–1501.

16. J. Terborgh et al., "Transitory States in Relaxing Ecosystems of Land Bridge Islands," in *Tropical Forest Fragments: Ecology, Management, and Conservation of Fragmented Communities,* ed. W. F. Laurance and R. O. Bierregard Jr. (Chicago: University of Chicago Press, 1997), pp. 256–274.

17. Ibid.

18. J. S. Coleman and S. A. Temple, "Rural Residents' Free-Ranging Domestic Cats: A Survey," *Wildlife Society Bulletin* 21 (1993): 381–390.

19. M. Soulé et al., "Reconstructed Dynamics of Rapid Extinctions of Chaparral-Requiring Birds in Urban Habitat Islands," *Conservation Biology* 2 (1988): 75–92.

20. W. S. Alverson and D. M. Waller, "Forests Too Deer: Edge Effects in Northern Wisconsin," *Conservation Biology* 2 (1988): 348–358.

21. W. J. McShea, H. B. Underwood, and J. H. Rappole, eds., *The Science of Overabundance: Deer Ecology and Population Management* (Washington, D.C.: Smithsonian Institution Press, 1997).

Chapter 8

1. J. N. Abramovitz, *Taking a Stand: Cultivating a New Relationship with the World's Forests,* Worldwatch Paper No. 140 (Washington, D.C.: Worldwatch Institute, 1998).

2. A. T. Durning, *Saving the Forests: What Will It Take?* Worldwatch Paper No. 117 (Washington, D.C.: Worldwatch Institute, 1993).

3. N. P. Sharma, ed., *Managing the World's Forests: Looking for Balance between Conservation and Development* (Washington, D.C.: World Bank, 1992); World Bank, *The Forest Sector: A World Bank Policy Paper* (Washington, D.C.: World Bank, 1991).

4. C. Flavin, "The Legacy of Rio," in *State of the World 1997*, ed. L. Starke (New York: Norton, 1997), pp. 3–22; W. K. Stevens, "The State of the Earth: A Report Card," *New York Times*, June 17, 1997, pp. C1, C8.

5. World Resources Institute, *Tropical Forests: A Call to Action* (Washington, D.C.: World Resources Institute, 1985); N. Myers, *Conversion of Tropical Moist Forests* (Washington, D.C.: National Academy of Sciences, 1980).

6. J. Terborgh, *Tropical Deforestation* (Burlington, N.C.: Carolina Biological Supply Company, 1992).

7. N. Myers, *The Primary Source: Tropical Forests and Our Future* (New York: Norton, 1984).

8. E. Boo, *Ecotourism: The Potentials and Pitfalls* (Washington, D.C.: World Wildlife Fund, 1990).

9. C. M. Peters, A. H. Gentry, and R. O. Mendelsohn, "Valuation of an Amazonian Rainforest," *Nature* 339 (1989): 655–656; C. M. Peters, *Sustainable Harvest of Non-timber Plant Resources in Tropical Moist Forest: An Ecological Primer* (Washington, D.C.: World Wildlife Fund, Biodiversity Support Program, 1994).

10. N. R. de Graaf, *A Silvicultural System for Natural Regeneration of Tropical Rain Forest in Suriname* (Wageningen, Netherlands: Wageningen Agricultural University, 1986); G. S. Hartshorn, "Natural Forest Management by the Yanesha Forestry Cooperative in Peruvian Amazonia," in *Alternatives to Deforestation: Steps toward Sustainable Use of the Amazon Rain Forest*, ed. A. B. Anderson (New York: Columbia University Press, 1990), pp. 128–138; F. Mergen and J. R. Vincent, eds., *Natural Management of Tropical Moist Forests* (New Haven, Conn.: Yale University Press, 1987).

11. M. Balick and R. Mendelsohn, "Assessing the Economic Value of Traditional Medicines from Tropical Forests," *Conservation Biology* 6 (1992): 128–130; T. J. Hodson, F. Englander, and H. O'Keefe, "Rain Forest Preservation, Markets, and Medicinal Plants: Issues of Property Rights and Present Value," *Conservation Biology* 9 (1995): 1319–1321.

12. K. Bouton, "Nobel Pair," *New York Times Magazine*, January 29, 1989, sec. 6, pp. 28–29, 60, 82, 86–88.

13. R. D. Simpson and R. A. Sedjo, *Valuation of Biodiversity for Use in New Product Research in a Model of Sequential Search,* Resources for the Future Discussion Paper No. 96-27 (Washington, D.C.: Resources for the Future, 1996).

14. K. Lindberg, *Policies for Maximizing Nature Tourism's Ecological and Economic Benefits* (Washington, D.C.: World Resources Institute, 1991).

15. IUCN (World Conservation Union), *Protected Areas of the World: A Review of National Systems,* vol. 4, *Nearctic and Neotropical* (Gland, Switzerland: IUCN, 1992).

16. N. Salafsky et al., "Can Extractive Reserves Save the Rain Forest? An Ecological and Socioeconomic Comparison of Nontimber Forest Product Extraction Systems in Petén, Guatemala, and West Kalimantan, Indonesia," *Conservation Biology* 7 (1993): 39–52.

17. de Graaf, *A Silvicultural System;* Hartshorn, "Natural Forest Management"; Mergen and Vincent, *Tropical Moist Forests.*

18. S. Schwartzman, "Extractive Reserves: The Rubber Tappers' Strategy for Sustainable Use of the Amazon Rainforest," in *Fragile Lands of Latin America: Strategies for Sustainable Development,* ed. J. O. Browder (Boulder, Colo.: Westview Press, 1989).

19. D. Podesta, "Slain Brazilian Ecologist's Legacy: Disarray in the Amazon," *Washington Post,* November 25, 1993, pp. A47–A48.

20. IUCN, *Protected Areas,* vol. 4, *Nearctic and Neotropical.*

21. P. M. Fearnside, *Human Carrying Capacity of the Brazilian Rainforest* (New York: Columbia University Press, 1986).

22. H. T. Tang, "Problems and Strategies for Regenerating Dipterocarp Forests in Malaysia," in Mergen and Vincent, *Tropical Moist Forests.*

23. L. C. Nwoboshi, "Regeneration Success of Natural Management, Enrichment Planting, and Plantations of Native Species in West Africa," in Mergen and Vincent, *Tropical Moist Forests.*

24. de Graaf, *A Silvicultural System.*

25. Hartshorn, "Natural Forest Management."

26. Nwoboshi, "Regeneration Success," p. 80.

27. T. C. Whitmore and J. N. M. Silva, "Brazilian Rain Forest Timbers Are Mostly Very Dense," *Commonwealth Forest Review* 69 (1990): 87–90; T. C. Whitmore, *An Introduction to Tropical Rain Forests* (New York: Oxford University Press, 1990), p. 176.

28. S. Edwards, "Metapopulations and Conservation," in *Large Scale Ecology and Conservation Biology*, ed. P. J. Edwards, R. M. May, and N. R. Webb (Oxford, England: Blackwell Scientific, 1993).

Chapter 9

1. N. Johnson and B. Cabarle, *Surviving the Cut: Natural Forest Management in the Humid Tropics* (Washington, D.C.: World Resources Institute, 1993).

2. R. Kaplan, "The Coming Anarchy," *Atlantic Monthly*, February 1994, pp. 44–76.

3. J. P. Holdren, G. C. Daily, and P. R. Ehrlich, "The Meaning of Sustainability: Biogeophysical Aspects," in *Defining and Measuring Sustainability: The Biogeophysical Foundations* (Washington, D.C.: World Bank, 1995), pp. 3–17; H. E. Daly, "Operational Principles for Sustainable Development," *Earth Ethics*, summer 1991, pp. 6–7.

4. E. H. Clark II, J. A. Haverkamp, and W. Chapman, *Eroding Soils: The Off-Farm Impacts* (Washington, D.C.: Conservation Foundation, 1985).

5. L. R. Brown and E. C. Wolf, *Soil Erosion: Quiet Crisis in the World Economy*, Worldwatch Paper No. 60 (Washington, D.C.: Worldwatch Institute, 1984).

6. Clark, Haverkamp, and Chapman, *Eroding Soils*.

7. K. D. Holl and J. Cairns Jr., "Vegetational Community Development on Reclaimed Coal Surface Mines in Virginia," *Bulletin of the Torrey Botanical Club* 121, no. 4 (1994): 327–337.

8. R. G. Healy, *Land Use and the States* (Baltimore: Johns Hopkins University Press, 1979); R. G. Healy, *Competition for Land in the American South: Agriculture, Human Settlement, and the Environment* (Washington, D.C.: Conservation Foundation, 1985).

9. D. M. Roodman, *Paying the Piper: Subsidies, Politics, and the Environment*, Worldwatch Paper No. 133 (Washington, D.C.: Worldwatch Institute, 1996).

10. N. Myers, *The Primary Source: Tropical Forests and Our Future* (New York: Norton, 1984).

11. J. Terborgh and C. P. van Schaik, "Minimizing Species Loss: The Imperative of Protection," in *Last Stand: Protected Areas and the Defense of Trop-*

ical Biodiversity, ed. R. Kramer, C. P. van Schaik, and J. Johnson (New York: Oxford University Press, 1997), pp. 15–35.

12. S. Postel, "Carrying Capacity: Earth's Bottom Line," in *State of the World 1994*, ed. L. Starke (New York: Norton, 1994), pp. 3–21.

13. L. R. Brown and H. Kane, *Full House: Reassessing the Earth's Population Carrying Capacity* (Washington, D.C.: Worldwatch Institute, 1994).

14. G. Hardin, "The Tragedy of the Commons," *Science* 162 (1968): 1243–1248.

15. G. Gardner, *Shrinking Fields: Cropland Loss in a World of Eight Billion*, Worldwatch Paper No. 131 (Washington, D.C.: Worldwatch Institute, 1996).

16. Ibid.

17. Brown and Kane, *Full House*.

18. Ibid.

19. Gardner, *Shrinking Fields*.

20. Roodman, *Paying the Piper*.

21. Johnson and Cabarle, *Surviving the Cut*.

22. Ibid.

23. C. V. Barber, N. C. Johnson, and E. Hafild, *Breaking the Logjam: Obstacles to Forest Policy Reform in Indonesia and the United States* (Washington, D.C.: World Resources Institute, 1994).

24. Ibid.

25. Ibid.

26. Ibid.

27. G. Graham, "State/Provincial Reports: Texas," *Natural Areas Journal* 16 (1996): 158.

28. A. P. Dobson, J. P. Rodriguez, W. M. Roberts, and D. S. Wilcove, "Geographic Distribution of Endangered Species in the United States," *Science* 275 (1997): 550–553.

29. L. E. Dwyer, D. D. Murphy, and P. R. Ehrlich, "Property Rights Case Law and the Challenge to the Endangered Species Act," *Conservation Biology* 9 (1995): 725–741.

30. T. W. Lippman, "Christopher Puts Environment at Top of Diplomatic Agenda," *Washington Post*, April 15, 1996, p. A10.

31. H. W. French, "Donors of Foreign Aid Have Second Thoughts," *New*

York Times, April 7, 1996, p. E5; "Can Foreign Aid Work?" *Washington Post,* February 14, 1997, p. A20.

32. H. E. Daly, "Operational Principles for Sustainable Development," *Earth Ethics,* summer 1991, pp. 6–7.

Chapter 10

1. M. Wells and K. Brandon, *People and Parks: Linking Protected Area Management with Local Communities* (Washington, D.C.: World Bank, 1992).
2. Ibid.

Chapter 11

1. A. T. Durning, *Saving the Forests: What Will It Take?* Worldwatch Paper No. 117 (Washington, D.C.: Worldwatch Institute, 1993).
2. L. R. Brown and H. Kane, *Full House: Reassessing the Earth's Population Carrying Capacity* (Washington, D.C.: Worldwatch Institute, 1994).
3. J. N. Abranovitz, *Imperiled Waters, Impoverished Future: The Decline of Freshwater Ecosystems,* Worldwatch Paper No. 128 (Washington, D.C.: Worldwatch Institute, 1996).
4. S. L. Postel, G. C. Daily, and P. R. Ehrlich, "Human Appropriation of Renewable Fresh Water," *Science* 271 (1996): 785–788; S. Postel, *Dividing the Waters: Food Security, Ecosystem Health, and the New Politics of Scarcity,* Worldwatch Paper No. 132 (Washington, D.C.: Worldwatch Institute, 1996).
5. R. Kaplan, "The Coming Anarchy," *Atlantic Monthly,* February 1994, pp. 44–76.
6. D. L. Hawksworth and M. T. Kalin-Arroyo, coords., "Magnitude and Distribution of Biodiversity," in *Global Biodiversity Assessment,* ed. V. H. Heywood (Cambridge, England: United Nations Environment Programme, 1995), p. 750.
7. "On the Brink of Extinction," *WWF Conservation Issues* 2, no. 5 (1995): 3.
8. J. H. Cushman Jr., "Logging in Siberia Sets Off a Battle in the U.S.," *New York Times,* January 30, 1996, p. A3.
9. C. V. Barber, N. C. Johnson, and E. Hafild, *Breaking the Logjam: Obstacles to Forest Policy Reform in Indonesia and the United States* (Washington, D.C.: World Resources Institute, 1994).
10. A. Flurry, "Clayquot Mass Trials," *Amicus Journal* 17, no. 3 (1995): 46–47.

11. "Asian Loggers Move In on Guyana," *RAN Action Alert,* January 1994, p. 15.

12. F. Pearce, "France Swaps Debt for Rights to Tropical Timber," *New Scientist,* January 29, 1994, p. 7.

13. C. H. Dodson and A. H. Gentry, "Biological Extinction in Western Ecuador," *Annals of the Missouri Botanical Gardens* 78, no. 2 (1991): 273–295.

14. G. Hardin, *Living within Limits: Ecology, Economics, and Population Taboos* (New York: Oxford University Press, 1993); R. D. Kaplan, *The Ends of the Earth: A Journey at the Dawn of the Twenty-First Century* (New York: Random House, 1996); N. Myers, "Environmental Refugees in a Globally Warmed World," *BioScience* 43 (1993): 752–761; P. Ehrlich and A. Ehrlich, *Betrayal of Science and Reason: How Anti-Environmental Rhetoric Threatens Our Future* (Washington, D.C.: Island Press, 1996); T. W. Lippman, "Christopher Puts Environment at Top of Diplomatic Agenda," *Washington Post,* April 15, 1996, p. A10.

15. M. S. Serrill, "Famine: Hunger Stalks Ethiopia Once Again—and Aid Groups Fear the Worst," *Time,* December 21, 1987, pp. 34–38, 43.

16. Population Reference Bureau, personal communication, 1998.

17. Population Reference Bureau, *1996 World Population Data Sheet* (Washington, D.C.: Population Reference Bureau, 1996).

18. G. Hardin, *Living within Limits: Ecology, Economics, and Population Taboos* (New York: Oxford University Press, 1993).

19. J. M. Goshko, "Regional Conflicts Threaten 42 Million around World, U.S. Study Finds," *Washington Post,* April 5, 1996; "Ethnic Wars Multiply, U.S. Struggles to Meet the Challenge," *New York Times,* February 7, 1993.

20. N. Nusser, "Panama Jungles Fading Fast—Critics: Ravaging Stems from Invasion," *Atlanta Journal and Constitution,* December 30, 1990, p. A24.

21. "Good and Bad at Game," *Economist,* July 6, 1996, pp. 69–70; C. S. Wren, "Angola Rebels Accused of Elephant Kills," *New York Times,* November 21, 1989, pp. A3–A4.

22. P. Shenon, "How to Support a Rebellion: Now It's the Jungle That the Khmer Rouge Decimates," *New York Times Magazine,* February 7, 1993, sec. 4, p. 4; "Cambodia's Wood-Fired War," *Economist,* June 17, 1995, pp. 35–36.

23. D. Lorch, "Returning Tutsi Herders Add to Rwanda's Strains," *New York Times Magazine,* April 16, 1995, sec. 1, p. 3.

24. Kaplan, "The Coming Anarchy"; R. D. Kaplan, *The Ends of the Earth: A Journey at the Dawn of the Twenty-First Century* (New York: Random House, 1996).

25. A. Lewis, "The Price of Truth," *New York Times,* August 4, 1997, p. A15.

26. IUCN and WWF (World Conservation Union and World Wide Fund for Nature), "Assault on Papua New Guinea Biodiversity," *Arborvitae: The IUCN/WWF Forest Conservation Newsletter* 4 (1996): 10.

27. Population Reference Bureau, *1996 World Population.*

28. B. Crossette, "A Global Gauge of Greased Palms," *New York Times Magazine,* August 20, 1995, sec. 4, p. 3; B. Crossette, "Survey Ranks Nigeria as Most Corrupt Nation," *New York Times Magazine,* August 3, 1997, sec. 1, p. 3.

Chapter 12

1. World Commission on Environment and Development, *Our Common Future* (New York: Oxford University Press, 1987).

2. P. Ehrlich and A. Ehrlich, *Betrayal of Science and Reason: How Anti-Environmental Rhetoric Threatens Our Future* (Washington, D.C.: Island Press, 1996).

3. H. Kane, "Gap in Income Distribution Widening," in *Vital Signs 1997: The Environmental Trends That Are Shaping Our Future,* ed. L. R. Brown, M. Renner, and C. Flavin (New York: Norton, 1997), pp. 116–117.

4. B. Crossette, "A Global Gauge of Greased Palms," *New York Times Magazine,* August 20, 1995, sec. 4, p. 3.

5. "Pakistan's Dismal Cycle," *Economist,* November 9, 1996, pp. 19–20.

6. G. Ayittey, "The Mystery of Growth," *Economist,* June 15, 1996, p. 6.

7. P. Shyamsundar and R. A. Kramer, "Tropical Forest Protection: An Empirical Analysis of the Costs Borne by Local People," *Journal of Environmental Economics and Management* 31 (1996): 129–145.

8. "Wildlife Refuge System Faces Uneasy Future," *New York Times Magazine,* June 1, 1997, sec. 1, p. 21.

9. R. A. Mittermeier and I. A. Bowles, "Reforming the Approach of the Global Environmental Facility to Biodiversity Conservation," *Oryx* 28 (1994): 101–106.

10. H. W. French, "Donors of Foreign Aid Have Second Thoughts," *New York Times,* April 7, 1996, p. E5.

11. B. Crossette, "Foreign Aid Budget: Quick, How Much? Wrong," *New York Times,* February 27, 1995, p. A6.

12. U.S. Bureau of the Census, *Statistical Abstract of the United States 1996,* 116th ed. (Washington, D.C.: U.S. Bureau of the Census, 1996).

13. R. A. Forrest, "Japanese Aid and the Environment," *Ecologist* 21, no. 1 (1991): 21–32.

14. M. A. Boza, "Conservation in Action: Past, Present, and Future of the National Park System of Costa Rica," *Conservation Biology* 7, no. 2 (1993): 239–247.

15. The Nature Conservancy, "Acquisitions South of the Border," *International Update,* summer 1995, p. 1.

16. R. Repetto and M. Gillis, eds., *Public Policies and the Misuse of Forest Resources* (Cambridge, England: Cambridge University Press, 1988).

17. R. Mendelsohn and M. Balick, "Private Property and Rainforest Conservation," *Conservation Biology* 9, no. 5 (1995): 1322–1323.

18. A. P. Dobson, A. D. Bradshaw, and A. J. M. Baker, "Hopes for the Future: Restoration Ecology and Conservation Biology," *Science* 277 (1997): 515–522.

19. D. M. Roodman, *Paying the Piper: Subsidies, Politics, and the Environment,* Worldwatch Paper No. 133 (Washington, D.C.: Worldwatch Institute, 1996).

20. D. M. Roodman, *Getting the Signals Right: Tax Reform to Protect the Environment and the Economy,* Worldwatch Paper No. 134 (Washington, D.C.: Worldwatch Institute, 1997).

21. T. E. Lovejoy, "Lessons from a Small Country," *Washington Post,* April 22, 1997, p. A19.

22. C. V. Barber, N. C. Johnson, and E. Hafild, *Breaking the Logjam: Obstacles to Forest Policy Reform in Indonesia and the United States* (Washington, D.C.: World Resources Institute, 1994).

23. P. Weber, *Net Loss: Fish, Jobs, and the Marine Environment,* Worldwatch Paper No. 120 (Washington, D.C.: Worldwatch Institute, 1994).

INDEX